W9-AGY-077

DARK AT THE ROOTS

DARK AT THE ROOTS

A Memoir

Sarah Thyre

COUNTERPOINT

A Member of the
Perseus Books Group
New York

Published by Counterpoint Press,
A Member of the Perseus Books Group

Designed by Timm Bryson

Library of Congress Cataloging-in-Publication Data
Thyre, Sarah.
Dark at the roots : a Memoir / Sarah Thyre.
p. cm.
ISBN-13: 978-1-58243-359-2 (alk. paper)
ISBN-10: 1-58243-359-3 (alk. paper)
1. Thyre, Sarah–Childhood and youth. 2. Houma (La.)–Biography. I. Title.
CT275.T625A3 2007
976.3'41063092–dc22
[B]

2006037251

10 9 8 7 6 5 4 3 2 1

For Andy

I like liars.
Liars care enough to make the world
a more interesting place than it actually is.

—Sarah Thyre

Contents

Author's Note

Most of the events described in the book happened as related; others were compressed, expanded, or altered. All names, except for those of some of the author's immediate family, have been changed, and some of the individuals portrayed are composites.

1975

This Is What People Do

"Would the mother of Renée Thyre please come to the front of the store?"

The assistant manager of Venture stepped away from the microphone and wiped his mouth on his wrist. We were up in his managerial aerie, a skybox that looked out onto Kansas City's first superstore. His name tag read "Garry" and he was a little greasy at the temples, but I could still imagine being happily adopted by him.

"Have some more popcorn, Renée," Garry said, shaking the bag at me. "Don't worry, I'm sure your mother will be here soon—c'mon, gimme a little smile!"

I smiled up at him the way an adorable seven-year-old named Renée would, batting my eyelashes like Disney's Sleeping Beauty, Briar Rose. I reached for the popcorn daintily, and was just about to ask for something to drink when my mother rushed up the steps of the platform and shot through the half-door.

"Oh my god, Sarah! I have been looking all over this store for you! I was just having them cut the fabric and I look over and you're gone! They still haven't caught that kidnapper, you know!"

Garry took a little half-step between my mother and me, his feet in ballet's third position. I felt so safe.

"Ma'am, I think there's a misunderstanding? This is Renée. Renée Thyre?"

Mom threw back her head and burst into laughter. "Well, that's what she told *you*. Her name's not really Renée! It's *Sarah*! But she's always telling people it's Renée! For some reason she just loves that name, Renée. I don't know where she even heard it. Can you imagine?"

Garry swept his feet from third to fifth position and swiveled his hips away from me. In that swift, subtle shift of his body, he left my side—me, Renée, his future ward!—and went over to Mom's. Together at last, they regarded me with heartless, unchecked mirth. I half-expected Mom to get on the store microphone and say, "Hey everybody! That announcement a little while ago—the one about Renée Thyre? Well, *there is no Renée Thyre!* It's just Sarah Thyre. 'Kay?"

* * *

"You're always ruining my life!" I cried in the car on the way home. "Why did you have to go and TELL him?"

The whole thing would have been worth it if I knew that just one person, just Garry, lived the rest of his life believing my name was Renée.

We'd gone to Venture that morning because Mom was having Prayer Group at our house later that afternoon. She needed to pick up a few yards of cheap calico fabric to make some quickie cloth napkins. All ours were stained and the Prayer Group ladies might notice.

Prayer Group consisted of Mom, four or five of her friends from church, and a Catholic priest named Father Donatello. The purpose of the meeting seemed more social than worshipful. The first order of business was grousing about the damn hippies who had taken over St. Lucy's and how things were so much better before Vatican II, when priests said Mass in Latin and women wore doilies on their heads. Pamphlets were distributed, or bookmarks embossed with a holographic saint. Lunch, usually a gourmet casserole, was served, but merely picked at if eaten at all. Hard liquor was poured. Piety drifted into levity, and levity exploded into ribaldry. Dance steps were demonstrated; cigarettes were French inhaled. When one of us kids got hungry or injured, the party would break up. Everyone staggered out to a chorus of "Best prayer group ever! God love you!"

My mother was a nervous wreck whenever Prayer Group was at our place. About a week before, she'd launch her campaign for me and my three sisters to pitch in and clean up the house.

"How did it get so bad?" Mom would say, kicking a path through the thick blanket of clothes, toys, and assorted paper products on our floors.

So bad? It was never good. Our house was always a mess. Banks of detritus built up against the walls, as though blown there by gale-force winds. You'd need a rake, possibly a pickaxe, to make a dent in it.

The night before Prayer Group, Mom began the preparations for her signature classy entrée, eggplant parmigian. She sliced and salted and stacked layers of eggplant between clean cloth diapers and cast-iron skillets to "leech out the extra juice," as she so appetizingly put it. In the morning she assembled the dish, layering the now brownish eggplant, Hunt's tomato sauce, an entire green can of Kraft parmesan cheese, and liberal sprinklings of dried Italian seasoning. The whole thing was topped with long-ignored mozzarella slices from the back of the fridge, trapezoid-shaped from having their moldy edges trimmed off. When she took it out of the oven, it had all coalesced into a rust-colored plank that drew rave reviews from adults, and suspicious "no thanks" from children. We knew a napkin-filler when we saw it. It was the perfect dish to serve when you wanted the kids to buzz off so you could live it up and pray.

When we got home, Mom laid the calico out on the dining room table and started to mark off squares with her fabric pencil. She looked up at the clock.

"Dammit, I won't have time to baste the edges. I'm just going to have to use my pinking shears."

The heavy, black-handled pinking shears would give the napkins a snazzy zigzag edge that, in a pinch, was an acceptable alternative to a finished seam.

"Where are my pinking shears? Sarah!"

I was the last one to get caught with the pinking shears. The day before, I had used them to trim my little sister Hannah's bangs into a nice jagged brim. Standing back to admire my styling skills, I nicknamed her "Cappie" on the spot. She burst into tears when she looked in the mirror.

Initially I played dumb. "What's wrong, Cappie?"

"I look stupid!" Hannah wailed.

"Aw, it's okay, Cappie," I said.

I couldn't even sell myself on this one. She did look terrible. Worse, she looked *poor.* Though Hannah's face was nicely tanned after a summer of playing outside, her new shelf of bangs revealed a strip of fresh white skin that underlined, practically illuminated, their crisp triangular edge.

"I don't have the pinking shears. You took them away and hid them, remember?" I said. An accusatory tone flavored my voice. Maybe Mom would think twice before stifling my budding creativity again.

She flared her nostrils. "Is that poddy? I smell poddy."

"Poddy" was our family word for the actual turd, the feces, the shit, the log. It never was "potty" or "pottie," and never meant the toilet/chair hybrid used for training toddlers. Nor was it used to describe the act of going, as in "Time to poddy" or, "Do you have to poddy?" It was used with "go," but poddy itself was always a noun. When one said, "I have to go poddy," one meant "I have to go *make* poddy," the way one would make cookies or an ice sculpture.

"Maybe it's Jeb's," I said.

Jeb was our spastic, semi-housebroken English setter. Technically, he was my father's dog, but Dad was often gone on business for three weeks at a time, so my sisters and I were forced to feed him and pat him on the head every once in a while. Eventually Becky and I would be caught mocking Jeb's dangling dogballs, and poor Jeb would be sent "to live on a farm."

"That doesn't smell like Jeb's poddy," Mom said, her eyes darting about. She spotted the offending item, a soiled cloth diaper in the corner of the room, shed by its wearer.

"Where's Debbie?" Mom said, looking around for my ten-month-old sister. As she picked up the diaper, several tiny turds fell out of it onto the floor.

"Ewwwww!" Hannah and I ran a safe distance away, into the hall that adjoined the dining area.

"Oh come on," Mom said, smiling like a genial TV neighbor. She clamped one of the pinky-sized pellets in the diaper and held it out as if offering a canapé. "Baby Ruth, anyone?"

I squatted down and sauntered over, dragging my knuckles on the ground. I took the diaper and held it under my protruding bottom, then pretended to throw it.

"Guess who I'm supposed to be?"

"Big Mac!" Mom and Hannah shouted.

Big Mac was the silverback male gorilla at the Kansas City Zoo. He was named after a popular menu item at his corporate patron, McDonald's. He was known far and wide for blindsiding spectators with handfuls of excrement.

That was too much for my oldest sister Becky, who had been sitting at the end of the dining room table, immersed in a Nancy Drew mystery.

"You guys're so gross," she said, slamming her book shut.

On her way out, she paused to inspect the turd. "What it really looks like is a Tootsie Roll."

Becky flounced out of the room in her indignant nine-year-old way, leaving me, Hannah, and Mom to a no-holds-barred "No wait, what does it really look like?" debate. Debbie crawled in, bare-bottomed. Grabbing the legs of a dining room chair, she pulled herself into a standing position and gurgled, pleased to see us so engrossed by the contents of her abandoned diaper.

It really broke the tension about Prayer Group. What did it matter if we put out the old stained napkins? There was poddy on the floor! Mom gathered up all her sewing tools into the uncut calico like a hobo's bindle and dumped it behind the couch. She sudsed the green shag carpet where the poddy had rested.

"Okay, enough," Mom announced in her company voice. "I'm going to go take out my curlers. Father Don'll be here any minute. Be polite when he says hello and then go play outside."

Our neighborhood in Kansas City was the one where newly married, lower-middle-class locals moved to get out of their childhood neighborhoods, which were becoming "dangerous." As though designed by Richard Scarry, the homes were faux Tudor style with identical three-bedroom floor plans, terraced front yards, and steep driveways. Our cousins, my mom's sister Barb and her family, lived two doors down. There were kids in almost every house on the block, with one house reserved for the requisite witchy old lady whose doorbell we rang and ran. We were the last of the Go Play Outside Generation. "Go play outside" was license to disappear all day long without anyone fearing for my safety.

My mother put out a stack of plates and silverware along with her eggplant concoction, a pitcher full of freshly mixed martinis, olives, toothpicks, and glasses.

Father Don arrived first, tenting his fingertips together at his lips and looking over Mom's shoulder at the spread on the dining room table. Back in the seventies, there was no shame in anyone, not even a priest, asking for a highball before lunch. In fact, priests were often the life of the party! They brought cases of altar wine to dinner and were least likely to cause a scene by getting into a drunken row with their spouse. Mom used to know a monsignor who kept a well-stocked bar in the trunk of his car in case he stumbled upon a party. Mom didn't approve. She had fallen out of touch with him, which I thought was a shame. He sounded pretty fun to me.

If Father Don was any fun, I wasn't around to see it. He was always sneaking up behind me, sliding his hand around my neck like a Crescent wrench, and whispering:

"Have you been honoring your mother and father? Pray the rosary every day."

Carrie Brower was next to arrive. She was tall, taller than my dad, who was six foot one. I couldn't tell if she was white or black. Her skin was sort of a greenish taupe and she wore her hair swirled up into a giant bun. If she were black, that would've been a semi-scandal in our neighborhood. We weren't racists per se, but we were definitely of the lock-your-car-door-there's-a-man-with-an-Afro variety. I was pretty certain a race war was about to start any day, and often begged Mom to keep plenty of canned goods in the basement.

Carrie was loud and holy and you could tell she didn't like children much. She used the first person plural to make it seem like she was on your level, but it was always mildly insulting.

"Oh my, don't we look dirty today? We could sure use a bath, couldn't we!" she said, reeling back as if she smelled poddy. Not a hair of her Jiffy Pop chignon was out of place. "Well! No doubt Mommy told us it's time for the grown-ups to pray and we should go play outside! Isn't that right?"

"I don't say 'Mommy' anymore," I grunted, feeling as dirty as she said we were.

Two other women I didn't know came in. Everyone stood around saying nothing, just sort of humming and smiling at each other awkwardly. Father Don rubbed his chin and said, "Ah, why don't we get started?" They all broke for the martini pitcher.

My sisters and I went outside and rustled up some neighbor kids for a game of Sardines. Sardines was sort of an inverted Hide 'n' Seek, wherein the person who was It hid, and everyone else counted to a hundred. The group split up to look for the hiding person, and as we found him we would quietly join him in his hiding place, cramming and crowding in together. The last person to find the hidden folk would be the next It.

We had a pretty good-sized group for Sardines. Mia and Pia Moretti had the most street cred because they went to public school and were known to be "wild." Mia had been suspended for setting her desk on fire and Pia had been picked up across the state line with a *boy*. Last year, my sister Hannah had stolen a necklace of Pia's, broken it, and shoved a bead from it up her own nose. Old Lady Hoover next door tried to get it out with a crochet hook but it was jammed so far up there, it just rolled around like the ball in a roll-on deodorant. Mom had to take Hannah to the hospital to get the bead sucked out with a giant vacuum. There was some grumbling over who should pay the doctor bills, and Pia's mom had grounded her. Since then the Morettis had a snotty attitude toward our whole family.

There was Mindy Nickles, who was Becky's age but more my friend. In the summertime, we shared popsicles with her dog Cindy, whose name I thought was "Sinny" because that's how Mindy pronounced it.

Mindy was the youngest of four, and they were all up for Sardines that day. Her brothers were Foxy Dan and Mean Pat, and her oldest sister Beth babysat for us on the rare occasions my parents went out.

Foxy Dan was twelve that summer and the alpha male of our street. When we reenacted scenes from *Walking Tall,* he always played the vigilante sheriff Buford Pusser, busting up our pretend white slavery den with a plastic neon baseball bat. When we played any hide or chase games, he was It first because he liked being It.

Whenever Dan was It, I tried my best to be the first to find him. I loved sliding in next to him in his hiding place, real close so I could see the blackheads on his nose. I loved the smell of him. He had that sort of sweet boy b.o.: green onion-y with just a tincture of rancid.

I was the youngest of the gang, and I wasn't very good at finding anyone at all, much less first. It wasn't long before I was It. Everyone hated when I was It because I was small and flexible and could tuck myself into a bathroom cabinet or basement window well and they would be forced to cram themselves in around me.

"Don't hide someplace tiny, Shrimp-tard," Mean Pat said, pinching me hard on the arm.

I hated Pat. Pat was the one who had arranged for me to fight Skippy a few weeks before. Skippy was five, the youngest of eight boys. He lived a few blocks over, on a not-yet-gentrified street. He rode around on his brother's yellow ten-speed with his long hair blowing in the breeze. Skippy thought he was hot stuff.

But I was seven. I was sure I could take him. So it was arranged. Skippy and I were corralled in a circle of teens and preteens. My cousin Louis officiated. When he said, "1,2,3 . . . FIGHT!" I moved forward just in time for Skippy to shove his hands into my cheekbones like my face was a volleyball. I crumpled. I cried. I went home, threw myself on the kitchen floor, and sobbed, "Mom! Call the police!"

"Skippy pushed you? That's not reason enough to call the police," she said. "Besides, isn't he only five?"

I put a wet washcloth to my face, refused dinner, and sobbed through *Petticoat Junction,* wishing I had a gruff, kindly Uncle Joe like the spunky girls on that show.

I was still mad at Pat for my public humiliation. This time I would find a good place to hide. I'd show him.

When we first moved into this house, the driveway ran up alongside and ended in a broad concrete slab where a garage used to be. Dad built a new garage, a soaring brown and yellow structure that felt like a barn. While it was under construction, evenings after dinner I'd sneak into it just to be alone. One night I went in there barefoot and punctured my foot on a giant bolt in the floor, earning the first of many lectures about lockjaw from my mother.

"Step on a rusty nail and you're dead of rigor mortis before you get to the hospital," she said. "I tell you one thing: I'd rather be dead than have to eat all my meals through a straw for the rest of my life."

Well, I had my shoes on today. Though brightly lit, the garage always felt a little spooky. It was my dad's place, and entering it without him felt illegal. It was filled, literally to the rafters, with junk, mostly his. A long workbench with a vise bolted to one end and a device for filling shotgun shells on the other took up one wall. A row of large galvanized aluminum chests filled with tools, extension cords, guns, and fishing equipment lined the other. I had recurring dreams that my family was being held hostage in the garage by a gang of Indians. In one, the Indians chopped my father up into tiny bits, diced but still in his original shape on the floor. In another, they carved a hollow in his back and made vegetable soup in it, stirring with a large paddle. The Indians let me take a turn stirring, until my sisters had to horn in on it and fight over who got the next turn, grabbing the paddle out of my hands.

Creepy as it was, I couldn't resist the wealth of new hiding places the garage offered. The small space left between the aluminum chests and the wall was a perfect place to hide. It was so narrow no one would think to look there. But if they did, the chests were up on raised, wheeled dollies so they could push them out a little to fit behind with me.

When the group hid their eyes and started counting, I raced to my premeditated spot. I slipped behind the chest, giving it a lick for good luck. There was something about the flavor of metal that was both bracing and relaxing. I used to lick a metal crucifix necklace I had, too. The hole at the top of the cross where the chain went through tasted spicy.

I counted silently to a hundred and waited to hear the "Ready or not, here we come!"

I didn't hear anything. I waited some more. I was a fast counter, they were probably still in the eighties or something. Still nothing. I licked the metal chest a couple more times, started to sweat.

I peeked out over the top of the chest. Nobody there. My eyes glanced up to the rafters, which were stuffed full of all the things people keep that they will never use again. In plain sight were the components of our old baby crib, made of dark brown particleboard decorated with big-eyed teddy bear decals. One of the solid end panels had an angry fisthole where my dad had punched through it. I was the one in the crib at the time.

Wait, what was that? I ducked back down. I thought I heard someone.

I peeked again. The door to the garage was still closed. What if it was a rat? I wasn't afraid of bugs. Bugs were cool. But the idea of something with a heart, a closed circulatory system filled with warm blood, brushing up against me, made me panic.

Great. Now I had to go to the bathroom. Well, it was just pee. It would dry up. I eased into a half-squat, worked my pants down around my ankles, and started to go on the garage floor. But as so often happens, with no advance warning, some poddy came out too. A not-insubstantial piece of poddy.

I was just pulling up my pants and hadn't given a thought to how the rest of the kids would join me in my now-sullied hideout when I heard the garage door open. The soft *fillop fillop* of thongs on the cement floor meant only one thing: Beth Nickles, the oldest of the gang, the wisest, the closest thing to a grown-up.

"Aw, crud," I thought.

I ducked my head down to peer out under the chest, holding my hair up to keep it from falling into my own waste. I saw Beth's feet and then her hair. She leaned down even farther, the neckline of her dashiki gaping to reveal her pointy new pacifier nipples. Her face came into view, breaking into a grin of "A-ha!" and then immediately into a look of pure revulsion.

"Man, you went to the bathroom! That's sick! I'm telling!" she said.

From somewhere deep inside me, somewhere primal and primeval, a noise emerged like a bubble. It rose up my throat and took the shape of a word, which then spilled out of my mouth. I was powerless to stop it, just as I had been with the poddy mere moments ago.

"*FUCK!*" I said.

"Oh my gosh," Beth said, doubly, triply horrified. "I am so telling right now!"

And with that she was gone, running out of the garage and into our house and straight up to my mom, tipsily ensconced in her Prayer Group. I raced after her, trying to appear rather moseying and unconcerned at the same time, and entered the living room just in time to hear Beth say, ". . . and then she said *the F-word*!"

Father Don, the ladies, and my mom looked at me, their buzz killed. They were too stunned to move. Mom leapt over the coffee table, banging her shin but not stopping to flinch.

"Young lady, I am *shocked* at your behavior!" she said, sounding like Miss Gulch in *The Wizard of Oz*. It was the same tone she took when I stole those Sweet Tarts from the drugstore and she made me return them and confess to the manager.

She dragged me toward the dining room.

"You're coming with me right this minute!"

"Where?" I wondered. I sure hoped we were going to get a snack. I was famished.

She stopped just outside the living room and said in a strange, high voice, "The F-word? That's filthy talk. We're going to wash your mouth out with soap."

"What the—?"

We heard a little moan. Hannah was sitting at the dining room table in front of the now-empty martini pitcher. She looked at us with bleary eyes, threw up a little, and fell off the chair onto the floor.

Without missing a beat, Mom marched me through the kitchen and into the half-bath at the back of the house. She picked up the much-used bar of Ivory from the counter, with its darkened crevices and layer of slime.

"What are you doing?!" I asked, aghast.

She never washed my mouth out with soap! I had heard rumors of this sort of punishment, but . . . of course I had never used the F-word before, either. I had never even HEARD it before. My parents were more of the "horseshit"/"god*dam*mit" school of cursing. That "*FUCK*" came from somewhere aboriginal within me, some vestigial remnant of DNA handed down from the first coelacanth that sprouted legs and decided to crawl up on land, only to find out terra firma sucks, too.

"Why? Why are you doing this?" I pleaded, for once not trying to be melodramatic.

Mom looked at the soap and then at the bathroom door, as if she didn't know why she was doing it. She paused for a split second, a tiny opening that had I been quick enough, savvy enough—I maybe possibly could have exploited to my advantage. Then she set her lips in a firm line, took hold of my neck, and yelled through the bathroom door, "I'm washing your mouth out with soap, young lady! Maybe this will teach you!"

Whew! I thought. *It was all an act!*

Mom felt my neck relax and looked at me sharply. I gave her a wink. I opened my mouth to give her one of those knowing clicks, the kind Burt Reynolds gave to pretty ladies or to kids who, against all odds, were gonna pull through.

She shoved the soap in.

I gagged, my arms flailing.

Mom leaned in close.

"Sarah, please. Just be quiet and let me," she said. "This is what people do."

Break Me Off a Piece o' Dat

I found the empty shotgun shell out in a puddle next to the street. Dad filled shells on a machine at his workbench back in the garage. I loved watching him: the buckshot, the cotton stuffing, the smell of gunpowder, the sharp click/crunch of the machine's lever, and the stern warnings never to touch any of it.

How'd this get way out by the street? I carried the shell away from my body between my fingertips, delivering it straight to my father.

"WHERE'D YOU FIND THIS?"

The volume, the menace, the out-of-nowhereness of his reaction were all pretty typical. Dad's response spectrum ranged from purple-faced rage to red-faced hysterical laughter. I was terrible at predicting which way he would blow.

"Your father's just moody," was Mom's stock explanation.

"I said, WHERE'D YOU FIND THIS?," Dad repeated, shaking the shell under my nose.

I panicked, remembering I wasn't allowed to go near the street.

"In a bush," I whispered.

"WHERE?"

Why oh why did I say bush?

"In a puddle," I said loudly, with a firm little nod.

"I THOUGHT YOU SAID YOU FOUND IT IN A BUSH," he screamed into my face.

"Well," I said, back to mumbling, "I found it in a puddle . . . under a bush?"

Dad clamped his hand onto the back of my neck and pulled me close, making sure I heard every word of his lengthy dissertation about my record of inconsistencies and my general, inborn tendency toward untruthfulness. How confused he was by my relentless dishonesty, by ME, how sickened, how "You must've gotten this from your mother." I started to feel sorry for him, having to deal with a no-goodnik such as myself. From here on out, I certainly wasn't to be trusted.

"You've always been the Liar in the Family," he said, brushing the Fritos crumbs off his lap and lighting up a Winston.

Thus began my career as the Family Liar. If something got broken, I was blamed. When things went missing, my dresser drawers were searched. Even when given a chance to deny any wrongdoing, I was still spanked.

One Sunday morning, I felt sick to my stomach. Mom gave me some Pepto-Bismol tablets and said, "Well, taking this medicine breaks your fast. You might as well stay home from church."

You're not supposed to eat or drink anything an hour before Mass, though Vatican II excepted medicine from this rule. Mom thought the old rules were holier.

Dad, who never went to church, was sitting at the kitchen table wearing his weekend clothes: a blue velour robe and purple bikini briefs.

"Get over here," he said.

I walked over to him, feeling like the Cowardly Lion approaching the Wizard of Oz. The walls throbbed green.

Glancing down at his bikini underwear, I noticed the very tippity-tip of his penis sticking out the fly. My eyes snapped up to the ceiling like they'd been struck with tiny baseball bats.

"Don't you roll your eyes at me, young lady," Dad said, grabbing my chin. "If you're so sick you can stay in bed all day!"

I took to my bed, listening to Mom threaten and cajole my sisters into their church clothes and out to the car. Dad went down the street to watch a football game with my Uncle Sonny. All of a sudden, I felt good enough to go spray the hose all over the front yard and sidewalk and myself.

By the time Mom and my sisters got home, I had completely recovered. I turned the hose on them, church clothes and all.

"Dammit, Sarah!" Mom said, shaking the water out of her ears. "Your father'll be back from Barb and Sonny's any minute."

My stomach started to hurt again. I ran upstairs and stripped down to my underwear and got back into bed.

The front door slammed. I heard Dad call out, "Why's the damn yard soaked?"

There was some murmuring below, then the sound of a grown man leaping up the stairs two or three at a time.

I burrowed under the covers, feigning sleep and bartering with God.

Lord, just let me disappear this one time and I will go to church no matter how sick I think I am!

The aroma of my Uncle Sonny's house preceded him up the stairs: my Aunt Barbara's perfume, White Shoulders. Their house reeked of it. Dad whipped the covers off me.

"Were you playing with the hose out front?"

"Hemmmm?" I yawned, stretching my arms up and rubbing the corners of my eyes.

My father grabbed my wrist and yanked me into a standing position. "Why's your underwear wet, Liar?"

Somehow, wet underwear makes a spanking hurt worse.

The job of Family Liar was a crummy one. It had gotten to the point where I didn't know whether or not I was lying anymore. If I was such a great Liar, couldn't I perhaps pretend to be something else? Yeah. Something pure and guileless.

At bedtime a few nights later, Mom read us Hans Christian Andersen's "The Little Match Girl." It's the story of a humble, blonde street urchin going door-to-door in the dead of winter, shod in slippers, trying to sell matches to rich people in their warm homes. No one buys her matches, and she's stuck out in the snow all night, lighting them one after the other to keep warm, until she runs out of matches and freezes to death. The spirit of her dead grandmother comes down and flies away to Heaven with her soul, but her body is found the next morning, New Year's Day. Frozen in the snow, she looks so beautiful and noble and pink-cheeked that the townspeople are positively smitten—alas, too late to buy any of her matches, but smitten nonetheless! *Smitten.*

Hey, I was a blonde. I was pink-cheeked. Or at least red-nosed: I had a big shiny scar on my nose from the time last winter when Mindy Nickles pushed me down a hill on my sled and my face smashed into the sidewalk. The scab was thick as orange peel and lasted for months.

How could I not be humble, going door-to-door like the Little Match Girl?

I didn't have anything to sell, but couldn't I use her methods to *get* something? I know—I'd ask people for food! I was hungry for all those cool snacks my mother would never buy: the ones that contained BHA, BHT, TBHQ, nitrates and nitrites and artificial colors and flavors. I'd use that line from the *Oliver Twist* TV movie: "Please sir, may I have some more?" I'd wear my turquoise Dearfoam slippers.

The next morning I started out early. I worked my way down one side of 71st Terrace and back up the other, trying to appear simultaneously cute and gaunt, knocking at doors and asking for something to eat. Folks parted most easily with a slice of cheese, which wasn't bad, but I preferred cold cereal, if they had the sugary kind. To spare them any trouble, I brought my own bowl, an old potpie tin.

I developed a talent for being in the right place at the right time. If one of my cousins was eating a pickle, I'd long aloud for something "green and crunchy." My cousins were always eating pickle chips with soy sauce and Lawry's Salt. Anyone would need a beverage after that. So, if someone popped open, say, a frosty cold bottle of Sprite, I'd position myself nearby.

"Did I just hear a cool, sparkling citrus mist? I bet there's something limon-y around here."

Before I knew it, my clever enterprise devolved into cheap stunts. I ate a Milkbone dog biscuit because someone told me it was a cookie. Once I ate my friend Lisa's booger because she claimed it tasted like a cherry Icee. It all snowballed at my seventh birthday party, when neighbors and relatives chipped in and got me a big box filled with all my favorites: raw hot dogs, American cheese singles, Snack Pack chocolate puddings, and little boxes of Fruity Pebbles.

"To Sarah," the card said. "Our Little Mooch!"

Mooch? Had I been mooching? As I understood it, a mooch has some of his own, but longs for the better, the tastier, the creamier, the saltier and more savory. Yes, we had plenty of that natural hippie cereal Heartland Crisp at home, but I wanted Cap'n Crunch. Was that coveting my neighbor's goods?

At last I knew shame. Not that kind of shame where you become aware of your behavior and know another path must be taken. I'm talking about Shame: The Mother of Invention! What I really needed was some cold hard cash of my own. Why should I have to be a slave to other people's tastes? There was more to life than pickles and cheese and Sprite and boogers. The lady who worked down at Friedson's Drugstore told me they were getting that new chocolate bubble gum in any day now.

I stapled the ends of an old belt to a shoe box and hung it around my neck like a cigarette girl. I filled it with old copies of *National Geographic*. Surely they would fetch a fair price around the neighborhood, especially the issue featuring Baby Hal, the baboon who was kidnapped and killed by his own aunt. For my gentler customers, I mixed in a few copies of *Maryknoll*, a missionary magazine filled with pictures of starving African children with distended bellies, gruel-smeared faces, and fly-encrusted eyes. Everything was priced to move at 25 cents apiece.

My first customer was Mr. Pennis, who lived alone in the black-shuttered house down on the corner.

"Would you like to buy a—" I began when he opened the door.

Mr. Pennis, bitter perhaps from a lifetime of teasing, told me to take a long walk off a short pier. He softened the blow by flicking a nickel at my head. As it ka-ching'd off my skull, I got a better idea.

I ditched the old magazines. I pasted a label on an old Folgers coffee can that said "Jerry's Kids." I sucked in my cheeks and put on a wan expression, as if the plight of those kids were draining away my own life essence. I threw in a little of my Jerry Lewis impersonation to make it seem official. I took off down the street with a noodle-legged swagger, just like Jerry in *Cinderfella.*

I'd start at Grandma Bacon's house. She wasn't *my* Grandma; she was just the neighborhood grandma. I had high hopes for her—she was generous. She always shared her cigarettes and beer, breaking everyone up with mysteriously amusing quips.

"I still got my playpen; there just aren't any toys in it!" she often said, playfully lifting her blouse up to show off a big purple scar.

"Good afternoon," I said when she opened her door. "I'm collecting for the telethon."

"Back again, Sarah? I'm all out of cheese," Grandma Bacon said, rubbing her face.

Something about her looked different—her eyebrows were gone!

I held out the can so she could read the label.

"Whuzzat—Jerry's Kids?"

"Oh, you know," I said. "They're the ones who talk real slow like this—'Hay-row, how yoooouuu todaaaaaaaay?'"

I crooked my arm and gnawed on my fingers a little, rolling my eyes like Jerry.

"Gaw-aw-aw-ayyyyy," I said, really getting into it.

"Stealing money from cripples and retards? Get out of here before I call your mother!"

Slam.

I'd have better luck with our next-door neighbor, Mr. Hoover. He was always generous, digging into his pockets to get me some Brach's Kentucky Mints, those powdery white candies with green jelly centers. They were a little stale and linty, but it was the thought that counted. A retired pharmacist, he brought over a big box of A&D Ointment samples every year at Christmastime. When my little sister Hannah lost her first tooth, Mr. Hoover told her if she didn't lick the space where it had been, a gold tooth would grow in. It wasn't his fault she didn't have enough self-control to keep her tongue out of there. His wife was the one who spent hours trying to fish that bead out of Hannah's nostril.

As usual, Mr. Hoover was sitting out in his front yard in a chaise lounge. Propped up next to him was the BB gun he used to shoot at stray dogs.

"Well, hello there, Blue Eyes," he said, all smiles. "Watcha got there, coffee can? We used to use those coffee cans to play—kick the can, or hide the can, or getcha two cans, punch coupla holes in 'em, take a piece of baling wire and run it through there, make a pair of stilts!"

I was pretty tired at this point and didn't much feel like giving him my Jerry's Kids pitch.

"Y'alright, Blue Eyes?" Mr. Hoover said, sitting up in his lounge chair.

I held out the can and rattled it a little. I'd put three pennies in there myself to give people the general idea.

"What's that?" His kindly, crinkly eyes widened into predatory saucers.

"Telethon?" I shook the can at him again.

"Nuh-uh, nosirree, little gal," Mr. Hoover said, hitching the chaise lounge into a more upright position. "No beggars on my property. Don't like dogs, don't like beggars. Not in my yard!"

In one smooth motion belying his old age, he shouldered the BB rifle and aimed the barrel at my chest, or at least that's what I planned to tell the police at my funeral.

I ran the twenty feet back to my own front door, tripping in my slippers.

Beggar? How dare he, that old Mr. Hoover! I was a salesman. Salesgirl. Okay, so maybe I was just a plain ol' mooch. This was no time for self-reflection. It was all the Little Match Girl's fault. She never had a BB gun pointed at her chest. Okay, she ended up worse. I could have taught that Little Match Girl a few tricks. She could have bundled up those matches with a snippet of decorative ribbon. She could have sweetened her sales pitch. When she stepped into the foyer of a potential customer, would it have killed her to say, "Mmmm, it's so cozy in here. Is that gingerbread I smell? Hey—nice chandelier!"

No.

Wouldn't have killed her.

* * *

Early in the summer of 1976, I escaped the pressing responsibilities that come with being the Neighborhood Mooch when we moved to Houma, Louisiana, the largest fishing village in the heart of the Mississippi Delta.

Houma's pronounced "HOME-ugh."

A curious thing happened to my father when we left Kansas City for the Deep South. He traded his peaked welder's cap for a high-crowned mesh hat with a CAT Power patch on it. He upped his tobacco intake, adding the occasional pouch of Red Man to his cigarette habit. He got rid of his teeny freshwater bass boat, buying the first of a string of muscular offshore fishing boats with 200+ horsepower outboard engines. He quit drinking scotch and water and started chugging beer. He changed his CB handle from Missoura Mule to Crawfish Tail. Most profanely, though, almost immediately he stopped using the second-person plural "you guys," replacing it with the foreign, inferior "y'all."

If it were possible to feel less close to him, we did. Mom, my sisters, and I watched from afar as he befriended people named Robichaux and Areneaux and Randazzo and began referring to himself as a "coon-ass." That's an insider's term of endearment among Louisianians of Cajun descent.

"Lay-zay lay bon ton roo-lay!" Dad would call out, firing up a giant pot of boiled crawfish out in the backyard.

One weekend, he went fishing on a tributary of a nearby bayou. He brought home six giant drumfish, four footers. They stank.

Dad dragged the fish out to the front lawn with a gaffing hook and hung them up on the fence.

"Sarah, get over here," he said, placing me close to one of the drum. It was long as I was tall.

He stood next to me, his teeth clenched into a smile. Mom snapped a Polaroid. Our new dog, an Irish setter named Sue, came up and sniffed the drum, her blood-spotted panties drooping. When Sue got her period, Mom put Debbie's old training underwear on her, with a hole cut out for her tail.

The neighbors, authentic Southerners who knew drum were garbage fish, strolled by, clucking their tongues.

"Y'all ain't gonna eat them river drum, are ya?" they said.

"An' why the hell we ain't?" Dad said, grinning flirtatiously at the local women and men alike.

He whipped out his fancy new electric carving knife.

"Looka dat Yankee knife!" hooted Pete Bruckner.

A round of chuckles convulsed the crowd. Grinning less widely, Dad hoisted one of the drum onto the top of an Igloo cooler. He plunged the knife in below the gills and started to work it down the body, flensing off a massive fillet the size of a winter scarf. When he peeled back the flesh, the fish's entrails wiggled with a startling degree of life, considering they'd been out of water for at least four hours.

"*Whoa!*" Dad said, bouncing backward.

The electric knife twisted around in his grip and sliced the tip of his ring finger off. Blood dripped down onto the squirming, worm-filled drumfish innards. Through their laughter, everyone tasted a little bile at the sight of it.

❋ ❋ ❋

My father seemed like an eager-to-please teenage girl, courting the popular Cajun kids. I watched and learned. Here we were in a brand-new town. If he could be a coon ass, I could pick something to be too. These people didn't know me from Adam.

I didn't want to be Renée anymore. Renées were a dime a dozen around these parts. A block down, on our very street, there was even a *guy* named René. He had tiny, shriveled legs and sat in a wheelchair with what appeared to be a bag of orange juice in his lap.

I was a precocious seven-and-three-quarters-year-old who immediately felt superior to the local hicks. They said things like "Y'all goin' t'tha thee-ATE-er or jest watchin' TEE-vee?" They pronounced cement

"SEE-mint" and of course there was the aforementioned "y'all," usually uttered with a circular, equine chewing motion of the lips.

I assumed the regal bearing of an intellectual, correcting my elders' grammar and dropping allusions to Nancy Drew's "Titian-colored hair" into casual conversations.

The Benoit family lived right across the street. They were half Cajun, half Texan. Hughlet Benoit was a year older than I, and Chantelle was a year younger.

Everyone in the neighborhood was always teasing me about liking Hughlet. I hated Hughlet. On the one hand, he did have nice blue eyes, but their allure was dimmed by a bowl haircut and his sharp little tetracycline-stained baby teeth. A nasty troublemaker, he would come over to our house, melt crayons onto the stove burners, crack eggs into the dishwasher, and put our Barbie dolls into the 69 position in their little swimming pool. He was the only eligible male on the block.

I probably had a little crush on him.

Chantelle was wiry and freckled. When she smiled her eyes turned into half-moons, and she was smiling most of the time, poor thing, just gazing into the void. She wasn't the brightest of bulbs, but she was highly suggestible and that suited my purposes just fine. Basically, she did whatever I wanted and believed everything I said. When I got bored with Chantelle and needed a bit more of a challenge, we'd go play a board game with the rest of the family.

During one friendly game of Junior Scrabble, Mrs. Benoit lay down tiles spelling out "P-U-D-L-E."

"Uh-huh," I said, my fingers curling over my tiles with equal parts derision and glee. "And by that do you mean puddle or poodle?"

"Why, *poodle*, sugar!" she said, laughing and pouring me more Barq's red cream soda. "Everybody knows you spell puddle P-U-T-T-E-L!"

"May I see the family dictionary, ma'am?" I began, anticipating the opportunity to provide supporting examples from Old French. I'd noticed a lot of words came from Old French when Becky and I played Fictionary with our friends. All you had to do to win was use the word "hence" and a lot of semicolons in your fake definitions. Next up: blowing these Benoits out of the water at Fictionary.

"Family dictionary?" said Mrs. Benoit, her tone souring. "What we got that's family is the family Bible and I believe that's just fine by us."

Oops—not as dumb as I thought. Smart enough to know I thought she was dumb. She escorted me from the premises, disinviting me from Chantelle's life. Slowly, the rest of the neighborhood shunned me. The occasions to fascinate, educate, and scintillate dried up. I looked forward to school starting so I could meet some fellow scholars.

How could I know that, education-wise, Louisiana in 1976 made Kansas City, Missouri, look positively avant-garde? St. Lucy's School in K.C. was progressive for a Catholic school. Grades 1 through 3 shared a huge basement room divided into "learning areas." We carried our books around in plastic "totes," storage bins with handles and sliding lids. There were no desks. We sat on the floor and used the flat tops of our totes for a writing surface. The teachers didn't talk down to us. My first-grade teacher, Mrs. Crouse, once told me, "Shut your trap, you, you, you–*beast!*" She made me feel like I was in a movie, or at least a play.

When we got to Louisiana, Mom took us to the schools to check them out. She was dead set against public school but still took us to look at them. I must've picked up on her vibe. All the kids in the public school classrooms seemed to hiss at us like caged psychopaths, strings of saliva connecting their overdeveloped canine teeth.

"This is more like it," Mom said, as we pulled up in front of St. Benen's, the preeminent old Catholic school in Houma.

In the school office, there was a fat girl lying on a dark green vinyl couch crying her eyes out.

"Right this way," said Sister Eustace, the vice principal, briskly leading us into her cold, austere office. She looked at us darkly and said, "Of course, if the girls don't behave, we *will* whip them."

Becky and Hannah and I immediately raised our hands to ask what specific behaviors would earn us a whipping, but Mom just swallowed hard and said, "I think you'll find my daughters probably won't need to be whipped."

She didn't sound so confident. Was whipping the same as spanking? It sounded much worse. With spanking you were laid over something: a knee, a bed, a couch. It hurt, but at least you could brace yourself.

Whipping sounded like you got hurled around the room, unless of course they tied you to a pole first. And couldn't it go on for, like, *hours?* That whipping song on the *Jesus Christ Superstar* album was interminable.

Our questions unanswered, my sisters and I were sent back to the outer room, where the fat girl was still sobbing. There was nowhere else to sit except for the vinyl couch, so we just stood next to it and stared at her.

"What're YOU lookin' at?" the girl said, seeming pretty tough in spite of her dripping nose.

"Nothing," we all said.

"Y'all gonna come to school here?" the fat girl asked, pinching a rope of snot between her fingers and pulling it away from her nostrils.

"Oh, probably not," I said.

"I'm Stephanie," she said. She leaned toward us and lowered her voice. "Sister Eustace'll whip your butt 'til it bleeds."

I stole a quick glance at her upper thighs, which were sticking out of her uniform skirt, splayed against the vinyl of the couch. I didn't see any blood, but one couldn't be too careful. I was already getting spanked plenty at home.

"We can't go to St. Benen's!" we whined in the car on the ride home. "They whip you there!"

"The kids at public school carry knives." Mom shook her head and shrugged sadly, as if it were out of her hands.

In the fall of 1976, I entered third grade at St. Benen's. My teacher, Mrs. Bouche, mispronounced all the Swedish names in *Pippi Longstocking* and claimed the North Pole was located in the country of Arctica. I wasn't even tempted to point out her errors. Not because I was scared of being whipped 'til my butt bled, but because for the first time in my life, I was the teacher's pet.

When the lunch bell rang, she'd wink at me and say, "Sarah, could you stay after class for a moment?" Then she'd give me and me alone a special little gift: a box of Kleenex, a little memo pad with a gorilla on it, a Hostess Ding-Dong. For no reason!

One day, Mrs. Bouche announced to the class that we were going to do nothing all day long but watch filmstrips and eat cupcakes in my honor, because I was representing our class in the school spelling bee that afternoon. For Christmas, she just happened to get my name as her Secret Santa. She gave me an inspirational plaque that read, "Friends Are Very Special People." On it was a picture of two little girls, one blonde like me and one brunette like her, holding hands and skipping down a flowery path under a rainbow.

Finally, I could take her aside and tell her there was no such place as Arctica.

During the last week of school before summer, there was a big canned food drive for the Campaign for Starving Children in Africa. The founder herself, Loubeth Markness, was coming to speak at our school. One student from each class would be designated a Hero for Hunger and would sit with her on the stage in the school auditorium during the assembly.

The Heroes would keep a log of what food their class brought in, and whichever class hated hunger the most won a hot fudge sundae party. Of course Mrs. Bouche picked me.

At the assembly, it was exciting to sit up on the stage with all the other Heroes. Mom let me wear a slip and suntan-color pantyhose under my uniform skirt.

St. Benen's principal came to the podium and introduced Loubeth Markness. She strode out onto the stage wearing a blinding white pantsuit. With her cinched waist, blonde flip, and upturned nose, she looked like Sweet Polly Purebred from the *Underdog* cartoons.

In a rich, seductive Southern voice, Loubeth Markness began to describe the dreadful famine in Ethiopia. She urged us all to be generous, to scour our cupboards, but not to bring in perishable items, and please, no more boxes of grits or fish fry.

"Donate things that you yourself would like to eat," she said.

Why on Earth would I do that?

Food drives meant one thing: bringing in a can of either beets or pumpkin. There were always several standing by in the dark recesses of our pantry. They were what Dad bought when he took the checkbook away from Mom and insisted on doing the grocery shopping himself.

"If you would just learn how to spend my money," he'd tell her, waving receipts in the air, "You wouldn't have to keep asking me for more!

These beets were eight for a dollar! Saltines—69 cents! And Jell-O—twelve boxes for a dollar!"

Loubeth went on, her voice beginning to quaver.

"The absolute best thing you can donate to Starving Children in Africa," she said, pausing to catch her breath, "is Chef Boy-Ar-Dee Ravioliss."

Ravioliss? Did she mean Chef Boy-Ar-Dee Ravioli? That was like, the Holy Grail at my house! I saved up my irregular allowance and the pennies I found in the couch for a can of that. I'd keep it hidden in a boot in the back of my closet until the coast was clear, and then I would heat it up on the stove and enjoy it. ALONE.

"Let me tell y'all a story," Loubeth went on, her eyes moist. "Last time I was over by Ethiopia, these little starving children came up to me, all lined up in a neat little line, and so hungry, bless their hearts. Well, I put a little ravioliss on their tongue, just one per child is all it takes, and they just sort of let it dissolve in there, like a lozenge, and one by one they all cracked a smile and said, 'Thank you, ma'am'—so polite! They just LOVE ravioliss! So please bring lots of that in. But please—only Chef Boy-Ar-Dee. The store brand has an off flavor."

Sitting on the dais behind her, I stared at the back of Loubeth Markness. If she wore that white pantsuit while she was dishing out the ravioli, she was asking for trouble.

The problem with being called a liar and a beggar and a mooch is that, even if you're fairly sure you're not *entirely* any of these things, there's always a little voice in your head that says, "Oh just go ahead and be that, just this one time. You know how."

When the canned food starting piling up in the box at the back of the classroom, I kept track of every item on a toteboard Mrs. Bouche had designed. "Sarah's Our Hunger HERO!!!!" it said in sparkly letters across the top.

I figured all that extra work earned me some ravioliss for my trouble. Alone in the classroom during recess, I was pulling two cans out of the box to sneak into my book bag when Mrs. Bouche came into the room.

"My, what generous students I have. And so much ravioliss! We really hate hunger, don't we, Sarah Sue-Bee?" She rubbed my shoulders and looked into the box. "Ewwwww, who brought the canned beets? Yuck!"

"Oh Butch, I know," I said, calling her my pet name for her. "Bleccc-cchhhhhh!"

She laughed and hugged me.

I pulled away, looking around the classroom to make sure the coast was clear.

"Dawn Whorley brought in the beets," I whispered. "And Butch . . . it's ravio*lee*. 'Kay?"

Smell It Like It Is

When we moved from Kansas City to Houma, my sisters and I rode with Dad in his green station wagon. The ride was dreary and silent, except for an occasional garbled blast from the CB radio.

"BRRRRRRRRT—TH—can I get a ten-thirty-six for that smokey over on Forty-Nine—THURRRRRRRRLLLLLLT" blared through the car, assaulting my ears. Where was the lyrical "Ten-four, good buddy," or the haunting "Big Ben, this here's Rubber Duck" from that chart-topping truckers' anthem, "Convoy"? The farther south we drove, the more incomprehensible and rambling the CB comments grew. Everything sounded like, "Weeellllllllll, ricky dicky doo dicky dicky roo, momback now!"

My sisters and I fought for a shift with Mom, who was driving alongside in her beige AMC Gremlin. When I got my turn, Mom let me keep her awake with loud AM radio hits like "Little Willy" and "50 Ways to Leave Your Lover." By the end of the three-day drive, she was even singing along with Paul Simon:

Just drop off the key, Lee / And set your-SELF free.

I couldn't believe my ears! *My* mother, singing along to contemporary, somewhat cool music, the lyrics all about getting the hell out of town to escape one's sinful-sounding "lover."

Mom was born in 1940. She came of age along with rock and roll, but she hated Elvis and the Beatles. In fact, aside from classical, she couldn't stand hardly any music at all.

"Turn that down! Music with words makes me feel like I'm going crazy," she'd yell, right when I was trying to tape-record "Bennie and the Jets" through the stereo speaker.

All my homemade cassettes were punctuated with similar complaints. The fade-out drum roll of "One Tin Soldier" was overdubbed with her muttering, "'Go ahead and hate your neighbor?' More like go ahead and hate this song!"

After three days on the road, we arrived at our new house late at night. A three-bedroom ranch in a treeless subdivision named Forest of Arden, it was completely unlike our tasteful, two-story 1920s urban Tudor back in Kansas City.

"Hey, look, everybody," Dad said with forced enthusiasm. "Instead of a garage, we're going to have a carport!"

"What's that?" I said, pointing to something out in the backyard.

In the middle of one of many standing puddles was a lone pole, a dirty ball hanging from it on a frayed piece of yellow rope.

"Tetherball!" Dad exclaimed, running toward the pole with his arm out to take a playful swipe at the ball. He slipped in the thick, primordial sludge at the puddle's edge and his legs slid out from under him.

"God fucking DAMmit!"

✵ ✵ ✵

That first night we slept in sleeping bags on the thick gold shag of the living room floor, the only carpeted room. The rest of the house had pebble-textured linoleum floors. I puked in my sleep: a thick peanut-butterish blurp. An omen.

There was vomit to spare in Houma. Maggie Blanchard, the girl in my class at St. Benen's with a monobrow and a voice that seemed to be struggling past a wad of oatmeal in the back of her throat, threw up something orange and chunky at least once a week. School bus bully Lacey Trosclair, with his poorly repaired cleft palate and luxuriously thick eyelashes, barfed during every First Friday mass, furthering the pastor's opinion that he was possessed by Satan. The cycle began in the school cafeteria, where a typical lunch consisted of a pile of disintegrating, unfamiliar beans up against a glop of fuchsia rice pudding.

"Probably red dye number six in there," Mom said when I complained about the school lunch. "I heard that ham they served last week was pesticide pork."

After a few months in Houma, Mom's quaint warnings about kidnappers were replaced with a new refrain: "You can get worms from that!" A surprising range of childhood activities could merit this admonishment: anything from a swift lick of raw cake batter out of the mixing bowl to walking barefoot on some moldy old carpet remnants in the drainage ditch out in front of our house.

Vomit. Worms. Peril was everywhere. Our route on the morning drive to St. Benen's took us through a long, dingy tunnel under the Intracoastal Waterway. Every day at the tunnel's entrance, Mom would say, "Ready! Set! Don't breathe, girls—this tunnel is filled with poisonous fumes that'll give you a brain bubble."

We clamped our mouths and pinched our noses shut for the next half-mile. If I couldn't make it through the entire length of the tunnel without inhaling, I seized up with nausea. My brain melted away inside my skull, and cancer sprouted in my lungs like those weird, Paleozoic-era ferns in our backyard. When I got to school, I'd mentally decide which of my classmates would be invited to my funeral.

"She was only eight!" the cute, blond Kenneth Poret would sob, clinging to my coffin.

Deep down, I knew I couldn't blame the tunnel. It was my fault I was going to get cancer. God would eventually get around to punishing me for looking at those dirty magazines.

Dad's *Playboys* and *Penthouses* were right out in the open, on a shelf of the short brown bookcase in the den. It was a humble piece of furniture, homemade in the early days of my parents' marriage. By this time it had a sway to it, with a pencil sharpener screwed into one side. An old globe with British Honduras and a single Germany was nailed to the top.

The magazines were sandwiched, appropriately and inappropriately, between the clinical, penis-filled *Life Cycle Library* and a ten-volume set of hardcover *Junior Classics*. Once Becky and I discovered them, every morning after breakfast we'd peruse them like the morning paper. We were careful to first lick the granulated sugar and French toast crumbs off our fingers. A tiny, bloody Jesus peeped over our shoulders from his crucifix on the wall above.

"Look at this one!" I said, opening the centerfold like a game-show host whipping the drape off a brand-new car.

There she was, the First-Naked-Lady-Who-Wasn't-My-Mom I ever saw: a bored-looking Japanese woman parting her crotch with well-manicured fingertips, staring off into space. Never before had I seen a real live Japanese person, a manicure, or that vacant expression.

What was she doing that *for?* My mother held open the pages of a book the same way, two fingers in a V, as if she were giving her other digits a rest. I felt sorry for this strange, slanty-eyed lady, with her tiny black nipples, her compulsion to hold open her hairy crotch, and a look that said, "Okay, *there*. Can I go now?"

Playboy quickly emerged as my favorite. The ladies had the prettiest hair. I figured they must use Body on Tap shampoo and Tame Cream Rinse. But even better, *Playboy* had cartoons. They always seemed to involve a Day-Glo-pink woman with baby-bottle nipples bending forward or presenting rearward or lying languidly under a lanky fellow wearing an ascot. Every italicized caption was something like, "*Are you sure you're NOT a contortionist?!?!*" I supposed it to be a worldly, adult sense of humor beyond my grasp, so I made a point to laugh extra heartily at the punch lines. To me, the cartoons were the real draw. Those tousled ladies with giant love beads hanging down between their tan boobs and fuzzy pubes sticking out like my Grandpa Eugene's moustache? BO-ring! But imagine, a *dirty cartoon?*

Looking at those magazines made my white Carter's undies damp. I thought it was pee. Even though I was looking directly at and sometimes into female genitalia, I didn't connect it with what I had going on down there myself. As far as I knew, I had one tiny hole at the tip of my "pee mountain." Poo came out the back from something that was not exactly a hole but more like a vortex, similar to those dilating aperture-like doorways in movies about the future.

The magazines were my gateway to noticing sex in general. One night when I was faking sick, Mom let me stay up late and watch *Barnaby Jones*.

"It's the Beverly Hillbilly's new show. Now he's a crime fighter."

This episode was all about some woman getting raped in the middle of a crowded shopping mall. The Beverly Hillbilly asked the woman what happened, and she said, "He made me take off all my clothes." Then she collapsed into hysterical sobs.

"What's rape?" I asked my mother.

"When a man makes a woman have sex," Mom said.

"What's sex?" I asked after a pause, somehow knowing I was treading on dangerous ground.

Silence. She had already bolted from the room. I concluded that sex was when someone made you take off all your clothes in the middle of a shopping mall. Later, when I found out about the actual mechanics, because of my pee mountain theory, I thought the penis was meant to go into my pee-hole. Horrified, I announced to my sisters, "I'm NEVER doing that!"

"You have to do it on your honeymoon," Becky said. "It's the law."

"Not me," I said. "If my husband loves me, he won't make me do that!"

One morning, Becky and I were caught looking. I was kneeling on all fours on the couch, arching my back and looking as blankly and slack-jawed as I could over one shoulder. I imagined a wire hanger hooked over my lower lip, pulling it down. Becky held up a *Playboy*, comparing my form with the centerfold's.

Mom came into the room and said what sounded like, "Caw! Caw!" She snatched the magazine away.

"Frickle frackle," she said, stomping out.

The delightful *Playboys*, the disturbing yet intriguing *Penthouses*: they disappeared for good. I mourned them. I missed them. Now I would have to make my own.

✵ ✵ ✵

Megan McClintock was my best friend in third grade at St. Benen's. Her regal-sounding name belied a lust for pornography as relentless as my own. During recess at school, Megan and I began to draw dirty pictures together. We did it in a three-subject spiral notebook with Lucy and Snoopy embossed on the cover, their feet blurred in that fast-stepping *Peanuts* happy dance.

Our scenarios were suave and sophisticated. Well-dressed, slitty-eyed men and women leered at each other from across fancy restaurants. In our drawings, exaggeratedly curvaceous, disaffected gals propped their vaginas open for crowds of wolf-headed, pop-eyed fellows, like a Warner Brothers cartoon directed by Tex Avery, only X-rated. In rooms that looked like the inside of *I Dream of Jeannie*'s bottle, men and women paired off into various sexual combos on plush Moroccan sofas. Cameras with *Playboy* logos tilted in on them from every corner of the wide-ruled page.

Megan was like a fashion editor, good at coming up with exotic locales and vaguely esoteric storylines. I was in charge of the humor page. I drew one cartoon of a worm looking down at his penis, saying "What the @#%$?" and another one based on that *Barnaby Jones* episode. A crying woman stood naked in front of a Sears store, saying, "Sex?!?! Wait! I thought this was rape!"

Jerry Blortans was the eldest of thirteen foster kids in a house down the block. He rode his bike around while wearing a form-fitting mask and skullcap he'd made from a popped plastic ball. He must've gotten his fashion-forward craftiness from his foster mom. Jerry told me she knit sweaters for his foster dad's penis. When I told Megan about this, we agreed it was an appropriate product to advertise in our magazine. I drew men with their hands on their hips, displaying their penis sweaters proudly, smiling and looking to the side like J.C. Penney slacks models.

When we needed inspiration, Megan and I would go to the Shamrock gas station out by the highway and check out the porn rack. I particularly liked the homey, black-and-white pictures in a magazine named *Stare*. The women were practically yawning, sitting in big rattan Papasan chairs with their legs spread wide. I would make a note of the art hanging on the walls or the knickknacks on the side table, like that clever figurine of a sleeping man wearing a sombrero.

Some of the women in *Stare* were photographed as they watched TV, chins on their fists, buttocks listing to the side. In one amazing picture, a lady with a big black Afro hiked her ankle up over a giant bowling trophy and managed to hold her crotch open almost *three inches* with only *one* finger.

Stare's combination of whimsy and laziness perfectly matched my own aesthetic. Megan and I decided to call our magazine *Stare*, too. In it, boobs looked like this,

and a penis looked like this:

The little penis sweaters were simple yarn cylinders that fastened with tiny buttons, which looked like this:

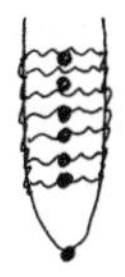

One weeknight during Easter break, I slept over at Megan's house. We went to see *The Shaggy D.A.* An old poster for *It's Alive*, the movie about a murderous, snarling, deformed baby, hung in the theater lobby.

On it was a picture of a bassinet with a menacing, bloody, baby-sized claw-hand hanging over the side.

That night in our zipped-together sleeping bags, Megan kept poking me with two curled fingers, saying, "It's alive, I tell you, alive!" so close to my face I couldn't escape her hot poo breath. Upon repetition, the stench made her teasing more annoying than funny, and I told her so.

"Back off, C.B.," I said.

"C.B.?" she said, still poking me and laughing.

I learned an important lesson that night. You can't call someone "Crap Breath" and expect her to keep drawing dirty pictures with you. Megan and I drifted apart. Actually, we parted more quickly, as so often happens when a couple shares a secret, shameful passion for filth.

A few weeks later, my parents announced it was "Spring Cleaning Day." My father liked to wake us up early on Saturday mornings by pulling the covers off us and tickling our feet, hard, until we got up to receive our orders.

"Today we're going to rake the yard!" he'd announce. "Put on your working smiles!"

Protests of "But Dad, it's the weekend." were ignored. We couldn't figure out his insistence on making us work on Saturdays until Becky went inside once to go to the bathroom and caught my parents having sex. Dad jumped up and ran into a closet. Mom sat up, pulling her t-shirt over her knees, and said, "It's how you were made."

Telling me about it, Becky cried and cried. I certainly never wanted to experience such horror. From then on, if I felt nature's call while out raking, I peed in the woods behind our house.

Today was an indoor job. Under close, non-copulating parental supervision, my face would ache from maintaining a working smile. My assignment was to clear out the junk packed under my bed: clothes, papers, shoes, paper clips, pennies, Life Savers, whatever I'd tossed below.

I started chipping away at the compacted block of refuse. Behind a small mountain of ossified snot rags and some dried-up Magic Markers, I came across the Snoopy and Lucy notebook that Megan and I had filled with our version of *Stare*. I leafed through it a little, chuckling at my jokes and admiring Megan's idea to draw a man balancing on his penis atop the spire of the Empire State Building: his arms gracefully outstretched, his eyes closed in bliss.

Ahh, those were the days! Well, won't be needing this anymore. Without a second thought, I toted it out to the trash can in my first armload of garbage.

When I brought out the next armload, my parents were crouched back in the corner of the laundry room next to the trash can, looking at the notebook. Mom put it behind her back real quick and they both looked at me, smiling and mad, like I'd caught *them* doing something wrong but I was going to pay for it. I could tell some kind of jig was up.

"Hi there!" I waved enthusiastically, then ran to hide out in my bedroom. I picked up a dirty sock and pretended to dust. Big beads of sweat broke out on my upper lip.

My parents came into my room quietly and closed the door. They sat on my bed and waited for me to turn around. I could feel their eyes drilling into my back. I continued to swipe the sock over the top of my dresser and whistle nonchalantly.

After a few minutes, my dad, wearing his trusty blue velour robe and purple bikini underwear, told me to turn around. Both of them were creepily calm and still sort of smiling. He held the notebook out as if he were making a presentation and said in his new Cajun accent:

"If I evah, EVAH catch yew drawing stuff like this again, ya better run for the hills, girl, y'hear?"

I shook with embarrassment. Not so much out of shame that they'd seen my intricately drawn sexcapades, but more out of fear that the tip of his penis was once again sticking out of his fly. Yes, it was 1976 and freedom was ringing, but why did Dad have to walk around the house dressed like Hugh Hefner's grandpa?

The possibility that his penis—which *Stare* had informed me was sometimes called a "dick"—was once again winking up at me . . . it was too terrifying to contemplate. I put all my energy into staring straight ahead. Dad's lecture about the evils of homemade pornography sounded like "blah blah nekkid people blah blah dirty times blah."

I kept my eyes averted so well, I was practically in a trance.

"Look at me when I talk to you!" Dad said.

I glanced at his fly.

Whew! No penis tip.

✲ ✲ ✲

When school let out that summer, Dad announced we would be making our first trip to Disney World. Mom checked out a book about Disney World from the library. My sisters and I pored over it, especially the section about accommodations. The pictures of the Disney Contemporary Hotel, a bold concrete flattened A-frame, showed the monorail shooting straight through the lobby. The gift shop had a huge gold mosaic mural on one wall. It all looked really fancy.

"Let's stay at the Contemporary!" I said, eager to abandon the hot choking clutter of our house for the hotel's clean-lined cement chill.

"No, let's stay at the one with the canopy beds and the Cinderella restaurant!" said Hannah, wearing a too-small chiffon dress backward with the buttons undone down to her navel. She picked up the hem of her dress and dabbed at the snot trickling from her nose.

"We're not staying in a hotel," Dad said. "I'm borrowing a camper from Darryl, a fella over at work. We're staying at Wonderful Wilderness Family Campground. A little roughing it would be good for y'all."

"Living off the land!" Mom exclaimed, rhapsodizing about crisp fresh air, piney woods, and babbling streams. She made the climate of central Florida in July sound like a deciduous Maine forest in late October.

Yes, yes, it would be just like *Little House on the Prairie*. I envisioned myself in a hand-sewn calico dress and homespun woolen bonnet, running barefoot like Laura Ingalls in *By the Shores of Silver Lake*, listening to my Pa play the fiddle while eating homemade doughnuts or a hearty bowl of popcorn and milk. The fact that I would also have buckteeth and freckles like Melissa Gilbert didn't bother me at all.

Dad brought the camper home a few days before our trip and parked it on the curb out front. I forgot all about the Contemporary Hotel and its monorail and gift shop. The neighbors came over to gawk until I agreed to give them a tour. *Noblesse oblige.*

"Yes, this is our new camper. My dad might let you borrow it sometime, I don't know. Now over here you'll notice the two-burner electric stove."

We left at three o'clock the next morning and the trip went downhill fast. Becky always got carsick, so she began puking right off the bat. The camper had a bad horizontal wiggle to it. It felt like an earthquake on wheels. Pretty soon we were all carsick. The toilet got stopped up and quit flushing.

"Eat a piece of bread to sop up the grease in your stomach," Mom said, bracing herself against the sway by wedging her feet under the built-in breakfast nook and putting her hands on the ceiling.

My mother's medical beliefs included an abbreviated update of the medieval theory of the four humours—blood, yellow bile, phlegm, and black bile—and their corresponding temperaments: happy, violent, dull, or lazy. In Mom's version, there were two humours: grease and mucus. A preponderance of either gave you a temperament prone to vomiting.

I staggered over to the kitchen area, looking for bread. When I opened the cabinets, canned goods fell off the shelves, pelting Hannah and Debbie and making them whine. Just as dawn was breaking, one of the tires went flat. We got towed to a service station outside Mobile, Alabama. Through the windshield, we watched Dad yelling at the mechanic and then into the pay phone. He swung up the steps of the camper and said, "The spare's BALD. They're charging TWENTY DOLLARS to put a temporary patch on the tire. HORSE shit. We're going home."

We drove back to Houma, the camper shuddering worse than ever on its patched tire. All of us lay in the back bedroom, by now dry heaving.

When we pulled up in front of the house, Dad jumped up and started maniacally flinging our luggage out of the camper and into the driveway.

"COME ON, LET'S GO IN THE STATION WAGON," he said, as the rest of us crept back onto solid ground, mute with cottonmouth.

That night we drove around Orlando for hours, looking in vain for a Vacancy sign. Around 1 AM, we found an empty parking lot. Mom spread out a blanket in the back of the station wagon.

"Lie down and go to sleep," she said.

My sisters and I sandwiched ourselves against each other. Mom and Dad sat in the front seat, passing a bottle of bourbon back and forth. They took short, sharp sips on it, like bites.

The next day we found a room at the Howard Johnson's in Kissimmee–St. Cloud, just outside Orlando. In their orange-and-turquoise dining room, under Dad's tabulating glower, I stuffed myself on fried clam strips and a frothy root beer float.

Sated, I completely forgot about reliving the wholesome hardships of Laura Ingalls and her prairie clan. We drove through the Wonderful Wilderness Family Campground just to see what we were missing, and it wasn't much. Lots of shirtless men scratching themselves next to stained tents, and rusty pickup trucks around a stagnant green "Kuntry Swimmin' Hole!" HoJo's had a sparkling chlorinated swimming pool. A beautiful body of water where, later that night, I met the man I was going to marry.

I was reaching for the ladder in the deep end when someone shoved me out of the way and climbed up ahead of me, the muscles in his tan legs snapping before my eyes.

"Todd, be nice to the little girl!" his mother called from her lounge chair across the pool.

Todd turned around and smirked down at me.

"Hey little *girl*," he said snidely, water dripping from his velvety brows.

He had to be at least twelve.

At Disney World the next day, I got a white sailor's cap at the Mad Hatter. I had them embroider "Sarah loves Todd" around the brim. Back at the hotel, I put on my best t-shirt. It was white with iron-on flowers and the words "SMELL IT LIKE IT IS" emblazoned across the chest. I tucked my hair up into my new hat and rushed out to the pool wearing it. There he was—that was him—dunking his little brother's head under the water.

I sashayed around the pool, swiveling my head back and forth so Todd could read the statement of my devotion.

"Hey, little girl!" he called.

I opened my mouth to form a shy, come-hithery "Hi, Todd," but he cut me off before I could get the words out.

"Hey, little girl, where's your boobs at?"

Back in our room, I watched myself cry in the mirror over the dresser and decided I looked best from the right.

✵ ✵ ✵

When we got home, the camper was still parked out front, listing to one side.

"When is Darryl coming to get that thing?" Mom asked.

"I don' know, woman. He's outta town on vacation hisself," Dad said, scraping his filet knife across an oiled whetstone.

The camper sat there for weeks, and it didn't go to waste. I took it over, made it my own personal clubhouse. A clubhouse that I could pretend to drive! Word spread around the neighborhood that this was the place to be.

The bedroom was in the rear, separated from the living/dining area by a ratty beige curtain. Wearing my "Sarah loves Todd" hat, I began coupling the kids off with each other and ordering them behind the curtain.

"Get in there and do it," I said to André LeBouef and Laura Chauvin.

Frightened, they looked at me and asked, "Do what?"

"You know what you have to do," I said to André, placing my hands paternally on his shoulders. "My boyfriend Todd, who lives in a big house with an Olympic-size swimming pool in Florida, told me you have to do it. It's the law."

I made the universal symbol for intercourse, sticking the index finger of my right hand through the "OK" symbol on my left. I nodded toward Laura.

"Rape her, André," I said, and pushed them though the curtain.

Everything was going great until one of the kids tattled. A woman down the street named Cherry found out and told my mother. The two of them cornered me in the bedroom of the camper, the scene of the crime.

"What you did here is nasty and sick," Mom said. "I'm going to have to pray the Lord forgives you."

"Listen up, Sarah, you better pray, too," Cherry said, trembling. "Pray I can get all this out of my head. Imagine: rape!"

Hours later, Mom was still hysterical.

"How can I show my face at church next Sunday? That motormouth Cherry in the choir," Mom cried, her face in her hands, sitting on the foot of my bed. "Pray for yourself, Sarah. Pray for all of us."

I prayed for amnesia to descend on our neighborhood. Something even better happened. Dad got transferred to a new job in New Orleans. We were moving to St. Tammany Parish, north of Lake Pontchartrain and eighty miles from this hellhole called Houma. I'd be leaving my pesky sex scandal eighty miles behind.

"Eighty miles—I'm sure they haven't heard a thing," Mom laughed and ruffled my hair fondly. "Just think: we'll be that much farther north, so it won't be so hot and humid! And your dad says we can join the country club!"

Country club?

A few months later, we were all piling into the car to drive to our new home in St. Tammany.

"Wait, I have to go to the bathroom!" I said, initiating my plan.

"Here's the key," Mom said. "Just pee? Don't wipe or flush. The water's turned off and Hannah had to go to the toilet, so the tank's empty. What will the new people think, moving in to an empty tank? You won't be able to wash your hands, either, so no finger foods for you."

I ran back inside and into the bathroom, locking the door behind me. I pulled the crumpled-up "SMELL IT LIKE IT IS" t-shirt out of my jeans, where I'd stuffed it earlier.

I took the lid off the back of the toilet and tossed the shirt into the empty tank.

Cheerio, darling! I thought, replacing the heavy porcelain lid. *I shan't be needing* you *at the country club.*

Git in the Truck

Piney River Country Club wasn't as fancy as Beau Chêne or Riverbend. It was more woodsy and rural, but it was still a country club. It had a golf course, tennis courts, a swimming pool with a snack bar, and a boat launch right on the thick, malted waters of the Tchefuncte River. Tchefuncte was the onomatopoeic name for the sound a big rock made—chuh-FUNK-tuh—when an Indian pushed it into the water.

We moved into our house on Tchefuncte Drive in the spring of 1977. It was another ranch house with linoleum floors in the kitchen and den; only this home boasted formal living and dining rooms, distinguished by their plush gold carpet. The four bedrooms were carpeted in a shade you could call Blood-Stained Watery Oatmeal.

Four bedrooms meant Becky and I each got our own room, with Hannah and Debbie sharing a bedroom, and my parents taking the master suite. My room was a narrow shotgun affair with a door on one end and a window on the other. The window came draped in a thick ecclesiastical red. I chose to paint the walls a rich pumpkin color, which made the room feel like the inside of a cornucopia.

We paid the cursory visit to the local public school. I made a point of checking out the school barn. Rumor was, Pickles and FayFay Mullit got caught having a "three-way" with a goat back there.

"The girls can't go to public school. The children there all have bowl haircuts. One of them had a port-wine stain covering his whole face," Mom said, lobbying my father for Catholic school tuition money.

Successfully, I guess. We ended up at St. Aldric's. Once again we were informed that we would be whipped if anything untoward arose, this time by the school secretary, a wide woman with eyes magnified by thick glasses.

"Ah'm not sayin' y'all *will* be whipped, I'm just sayin' y'all need to watch y'all's selves," she said, speaking into her shoulder as if there were a hidden microphone in her polyester lapel.

Mr. Lusco, the potbellied principal, had a comb-over from ear to ear and a deep, quarter-sized depression in the middle of his forehead. He played just a little too rough, bellowing and tossing kids into the air like rag dolls.

Lusco was the comic foil, a front for the real man in charge, the school pastor. A stern-faced Rod Steiger look-alike, Father Adrian wore thin nylon disco blouses while he mowed the school's lawn on a giant tractor. His stylish appearance concealed a darker side. By the end of my first day at Saint Aldric's, I'd been informed he was the one called in to administer the real discipline and that his weapon of choice was a black leather belt with a rusty nail duct-taped to the end of it.

Mr. Lusco didn't seem like the whipping type, but there was something unnerving about him nonetheless, like being around a grizzly bear that's trained to hug. I had my first run-in with him a week after I started school.

"Hey Thy-uh, whatcha doin' THY-uh?" he said loudly, grabbing me and throttling me.

How did he know my name already?

Lusco pulled off my acrylic ski cap, which had the Baby Ruth candy bar logo knitted into it, and threw it up into the air from hand to hand, playing keep-away with himself.

"Babe Ruth? Ya like Babe Ruth?" he taunted.

"It's BAB-Y Ruth, and it's a candy bar," I corrected him.

"Candy bah? Candy bah?" he said in that southern Louisiana accent that sounds like a Long Island accent. "Candy bah, candy bah, candy bah! BABE RUTH, THY-uh, BABE RUTH!"

It is an odd thing to be teased by one's principal, especially in sing-song. I retaliated by spreading a rumor about the origin of the dent in his forehead.

"You know how it happened? A butterfly kicked him in the head," I told everyone, stealing a line from the Paul Zindel novel *I Never Loved Your Mind.*

Dad had given me a boxed set of Paul Zindel books last Christmas, not knowing that they dealt with such subversive topics as high school abortion, parental dipsomania, getting high, and losing one's virginity to a transient hippie girl who ate macrobiotic cheese aged in goat dung. They had catchy, nonsensical titles like *Pardon Me, You're Stepping on My Eyeball!* and *My Darling, My Hamburger.* Dad had given me another book that Christmas: a little red manual about how masturbation and marijuana were equally bad and to be avoided at all costs. I didn't know what to think anymore.

With my flair for plagiarizing witty repartee, I attracted the friendship of Bitsy Marshall. She showed up on our doorstep a few days after I started at St. Aldric's, brandishing a tennis racket with a quilted blue gingham cover, looking for fresh blood.

"You nine?" she asked, throwing down the gauntlet. "Yeah," I said. "I mean, I'm eight and three-quarters."

"You play tennis?" Bitsy shot back.

"Sure," I said, already ashamed of my old wooden Wilson racket with two strings missing. "But all I've got is my old racket. The moving company lost my new one. It was silver."

"This one's graphite," she said. "My deddy got it for me because I learned to backhand one-handed last summer."

Bitsy led me on a secret trail through the woods that opened right onto the Piney River tennis courts.

"You like to play on rubico, or SEE-mint?" she said casually.

"Oh, ceMENT," I said, preferring on sight the neatness of the concrete, and taking the opportunity to lead by pronunciation example.

"Makes no difference to me," she said with a worldly air. "I could care less."

"You mean, you *couldn't* care less," I gently corrected. "Saying you *could* care less means you care *some*."

"Same difference," Bitsy said, wrinkling her nose.

I considered parsing that one, but feared my new best friend would take the moron part of oxymoron too personally.

"Look, it's that tennis pro who looks like Luke Skywalker!" she said, giggling like a girl at a sock hop.

I looked over and saw a blond, blue-eyed guy dressed all in white, volleying with a girl of about thirteen. Yes, there was a slight resemblance beneath the layers of sweat and acne and knee-high tube socks.

Turns out Bitsy and I had both just read Louise Fitzhugh's *Harriet the Spy*. Before long we were well on our way to becoming first-class voyeurs.

We built a fort in the hedge behind her house and spent hours spying on the old man next door. He wasn't just some innocent geezer sunning himself in boxer shorts and sheer nylon trouser socks, oh no. I was something of a self-styled maven on escaped Nazi war criminals. I had just seen *In Search of . . . the Angel of Death.* I was fairly certain this "Mr. Bergeron" was none other than Dr. Josef Mengele. Well, he wasn't going to inject blue dye into Bitsy's beautiful brown eyes or mistake us for twins or perform any of his other dastardly experiments.

We devised a clever trap, leaving some bait in his mailbox, a letter on Bitsy's mother's thick linen Crane's stationery:

Hi Josef,

We know you shot blue dye into people's eyes at Auschwitz. You ARE Dr. Mengele, alias THE ANGEL OF DEATH, AREN'T YOU?! You can tell us. Place your response in the ball washer near the coke machine at the clubhouse.

Respectfully, the Gestapo

We hid in the bushes and waited for the evil Angel to come check his mailbox.

"Wait!" Bitsy said. "That stationery has Mama's monogram on it!"

"Don't worry," I said, almost calling her "darling." "I'll go get it. I'll save us!"

I would've done anything for Bitsy. I loved everything about her: her pinched-in nose, her thick chestnut hair, her cowlick, her scoliosis, her jauntily cocked hips, and especially the two dimples above her butt on either side of her spine. I even loved that she said, "Those aren't dimples! Those're scars from two mosquito bites that I scratched and they bled and then I picked the scabs."

One day I put on my brown short-sleeved v-neck velour top, intending to seduce Bitsy on her mom's old lavender satin quilt up in the attic. First, we might sneak a nip of crème de menthe from the cut crystal decanter on the living room coffee table. Swiping the book from Mrs. Marshall's room to use as a script, we would reenact a scene or two from *Sybil.* Bitsy would be Sybil and I would play her mother.

"Sybil, want some milk and cookies?" I'd cackle, pretending to stick a flashlight in her vagina.

"Who dat who say who dat who say who dat?" Bitsy would answer. *"I'm not Vanessa."*

Then I'd turn into a handsome kidnapper and hold Bitsy hostage in a barn. Rescue helicopters would circle overhead, shining their searchlights (the flashlight again), but Bitsy would refuse to leave me. We would aim imaginary pistols at each other's heads and simultaneously pull the triggers.

Afterward, Bitsy's mama would make us Kraft macaroni and cheese with a whole stick of butter, and we'd smoke the crayon cigarettes we kept in a metal Band-Aid box. It was going to be a perfect day.

I walked through the woods alongside the golf course, taking the shortcut to Bitsy's house. Today I felt extra tough and outdoorsy. I pretended to be Jeremiah Johnson on the way to see my Indian squaw wife. I briefly wrestled with the question of how a luxurious satin comforter ended up in our wigwam. Oh well, I'd just pretend it was a grizzly pelt given to us by that crazy coot Bear Claw Chris, the guy in the Jeremiah Johnson movie who also played Grandpa Walton on TV.

"Huntin' grizz," I said to myself, picking my way along the trail.

I saw something sticking out of a bank of matted pine needles.

It was a stack of waterlogged magazines. Without touching them, I could immediately see genitalia hanging off their curled pages.

Dad didn't keep *Playboy* around the house anymore. Just as I had graduated from picture to chapter books, he'd moved on to dirty prose. *Between Phyllis and Sylvia* was sandwiched in the stack of jeans on his closet shelf, next to a loaded revolver in its leather holster. On the book's cover was the photo of a naked woman, presumably Phyllis, the bluebird tattoo on her hip peeping out from under a gloriously rumpled silk sheet. My hurried gleanings revealed that this tattoo was often lovingly traced by Sylvia's fingertips between moments of knuckle-clenching ecstasy.

I hastily buried the magazines so no one else would find them. I ran to get Bitsy.

"Hurry, let's go. Something to show you," I said, out of breath.

I propelled her back down the trail with my hand on the small of her back. When we got to the spot, I dug it out. I didn't need to say a word. Bitsy grabbed half the stack and we toted it deeper into the woods, behind a fallen tree.

"Sarah, wait." Bitsy dug the Band-Aid box out of her pocket. Instead of our usual crayons, inside were two real Pall Malls, from her mother's secret stash in their deep freeze. I never saw Mrs. Marshall smoke, but there was always a clean ashtray in her bathroom next to a bottle of Binaca spray.

Using a green Bic lighter, Bitsy lit us up two. We blew hard on the filters, making the cigarette tips glow. After a few tokes, we pounced.

The porn in the woods by the Piney River golf course was decidedly less genteel than Miladies Phyllis and Sylvia. The pages we managed to salvage chronicled a chance meeting between a large black construction worker, his occupation indicated by his muddy work boots, and a bored white housewife, her occupation connoted by a vanity table covered with cosmetics.

On the first page of their tale, "The Pearl Necklace," they were sharing a cup of coffee together in her kitchen. Next thing you know, WHAM! His big black beer-can penis is between her boobs, pointing right at her gold—not pearl, oddly enough—necklace. Her glossed lips had that wire hanger thing going on.

"Oh, I get it—*the pearl necklace,*" Bitsy said. "Get it?"

"Of course," I said.

I had a papery taste in my mouth. I lost interest in seducing Bitsy. One thing could lead to another, and then where would we be? I was utterly flat-chested, and so was she. Even if one of us had a penis, the pearl necklace would have nothing to grasp onto. It would slide right off.

This brush with porn brought out the girly-girl in Bitsy. She started wearing sundresses and espadrilles instead of jogging shorts and Tretorns. She tried to get everyone to call her Elizabeth Marie. She inspected the labels of my clothing, demanding to know their provenance.

"'Julie Girl'?" What kind of brand is *that*?" she asked, casting aside my strapless terry cloth romper from TG&Y, the local dime store.

"Wellllll," I said, my brain clicking and whirring.

"Where'd that come from, anyway?" Bitsy wanted to know.

"It's probably from Gus Mayer or Maison Blanche," I said airily, naming two high-class New Orleans department stores in which I had never set foot.

"Hmph, I never saw anything 'Julie Girl' there, or at Godchaux's or D. H. Holmes, either," she said, trumping me by ruling out the other two fancy stores across Lake Pontchartrain.

Bitsy walked over to our pantry and began to examine food labels.

"Thrifty Maid? Shur-fine? And what are these plain black-and-white boxes that say, 'soda crackers' on them? Don't you have Nabisco and Kraft and Campbell's? Hrmph. I'll just eat at home where we have real food."

After she left, I went into my room and flopped onto the bed, with its homemade spread. I played with a stack of tiny straw sombreros Bitsy and I had stolen off Torada tequila bottles down at the Winn-Dixie.

If the only kind of girl I could be was a Julie Girl, I might as well quit trying to be a girl at all. In fact, I'd ditch the whole country club scene entirely. Piney River wasn't all that la-di-da anyhow. Every summer there was a pink eye epidemic from the green, underchlorinated pool and a Hepatitis A outbreak from the box lunches at the Annual Ladies Golf Scramble. Miss Robichaux, the wealthy dowager who lived next to the boat launch and owned the land where the Piney River clubhouse sat, would shoot you with salt pellets if you tried to go fishing in the pond near her house. She'd come down to the country club pool and lie on a chaise lounge with her bikini top pulled down, exposing her droopy brown nipples. Everyone just smiled and pretended it wasn't happening, or that she was a sweet old lady with a European flair.

Yeah, well, I for one was tired of trying to keep up the charade. I wanted to swagger and curse like a boy, pull ticks off my balls with my bare hands, and jerk off my big black beer-can penis if I felt like it! I started to feel really butch. I needed a male role model, someone to look up to, maybe make out with.

The only guy I could think of was pimply Luke Skywalker over at the tennis courts. I practiced what I'd say to him:

> *Look, yes: I went as Darth Vader for Halloween. Sorry. I had to be Darth Vader, okay? The mask was marked down half-price and I already had the boots. Yeah, that was my little sister Hannah dressed like Mrs. Roper from Three's Company. I know: stupid! My mom made me take her trick-or-treating. She's so dumb she almost ate a homemade popcorn ball. Probably had a razor blade in it.*

Dang, that would never work.

Guess I'd have to turn to the only other man in my life.

✵ ✵ ✵

Back in Missouri, Dad shot skeet and occasionally went hunting. One night he brought home a dead rabbit. He skinned and butchered it in the kitchen sink. Mom deep-fried it like chicken and served it with boiled cabbage. When we sat down to dinner, Becky bit into a drumstick.

"Ow!" she said. "I think I got a piece of bone."

Patooey! She spit out a shiny BB onto her plate.

"My tooth! A piece of my tooth is missing!" she said, licking around the inside of her mouth. "I ate it! I ate a piece of my own tooth!"

"God DAMN," Dad said. "Lay off the rabbit."

He didn't have to tell me twice. I was already scheming on how to sneak the gamey riblets from plate to pocket to garbage can. Any money saved by eating hunted rodents was negligible compared to the cost of dental repairs. Or getting a piece of tooth surgically removed from one's stomach. I'd had a close call once, sucking on a nickel.

"You won't be happy until you have to get something cut out of you, will you?" Mom said, whacking me between the shoulder blades.

Since we'd moved to Louisiana, Dad had gone all-out with this fishing thing. Every weekend he loaded up his rods and tackle boxes, hitched the boat to the back of his truck, and headed out early Saturday morning to points south: Venice and Empire and Cocodrie and every other little town along the Mississippi River Delta. On Sunday afternoon he'd come home sunburned and stinky, lugging two or three Igloo coolers full of redfish and speckled trout, or sheepshead and amberjack, depending on whether he'd fished the marshes or the rigs.

Friday nights Dad started making the rounds, up and down the hallway, banging on bedroom doors.

"Who's going fishin' with me tomorrow?"

Hannah hid and Becky pretended to be asleep. Conveniently for her, Debbie was too young. But I started saying, "I will." I didn't particularly like fishing, but I felt too guilty to turn down his invitation when none of my sisters would go.

Soon I had a reputation as the Daughter Who Loved Fishing. You couldn't keep me from it! Going out on a boat with a bunch of crusty old Cajun guys gratified my new tomboy persona. I got to use tools. I broke up blocks of frozen bait shrimp with an ice pick and eviscerated redfish with a filet knife to get the swallowed hooks from their bellies. I learned to spit, whistle in that old-timey warbling way, and curse "da Nigras." I rubbed Coppertone cocoa butter into my skin and got so burnt, I'd peel three or four times. When dark brown freckles came out on my shoulders, I cut off the sleeves of my t-shirts to show them off.

Dad and his buddies shot the shit about politics and sports. I doused my ham and butter sandwiches with Tabasco and drank cream soda, eavesdropping on the man-talk. Its low, predictable grumble was soothing.

"Well, times are changin'," someone would always say.

"They can change all they want," my dad would always answer, "as long as they don't change the taste of Miller Lite."

Guffaws all around and swigs of agreement. A polite pause, then a grunt as one of them rose to a standing position.

"Gotta go wring out my sock."

Whoever said it would go to the stern of the boat, whip it out, and let loose with an arching cylinder of urine that hit the water sounding like a crisp carbonated beverage being poured over ice.

Bent over the picking box, sorting shrimp and crabs and squid with my hands in Dad's work gloves, I pretended not to watch. Boy, did that look like it felt good.

It wasn't fair. *I* had to squat on an old bait bucket, with nothing to wring, the sound of my urine a muffled *sshhuurrrlll*. Then on top of it all, I had to ask for toilet paper.

Oh, the shame, the shame of internal genitalia! As far as I could tell, a penis looked like the snout of some prehistoric animal. Yet I couldn't help admiring rather than pitying it.

On a three-day trip to Cat Island, Dad and I weathered an overnight storm on our twenty-two-foot Robalo. We slept on the bow, our bodies rising at a thirty-degree angle with each swell. The other guys called us crazy and slept in a tent on land. Oh, how I longed to set up house in a cozy little tent.

Dad and I went ashore at dawn to evacuate our shell-shocked bowels. E. J. Waguespack spat on the sand and informed my father that I was bad luck.

"Dat goil o' yours is makin' God kick up a storm and ruinin' our fishing," he drawled. "Fish don't like goils. Dey can smell 'em."

I hate you, E. J. Waguespack! I felt the way I did when that old man at St. Aldric's ice cream social kept calling me "sonny."

"Hands off, sonny!" he said as I reached to play the xylophone. The all-nun band was taking a break.

"Sister Bart told me I could play it," I said in a tiny, possibly inaudible voice.

"No-ho-ho, sonny," the smelly old geezer said, removing the sticks from my hands and slapping my fingers.

Just because I was wearing a purple t-shirt and green jeans and secretly coveted a penis didn't mean I liked being mistaken for a boy.

I couldn't say a damn thing to E.J. because my dad would say, "Don't you give him none of your lip, girl,"–or, in all likelihood, present company considered–"goil." He might even give me a smack. Old Cajun guys weren't known as anti-spanking advocates.

All day long, E.J. rode my dad about my fish-spooking stench. By lunchtime, no one had caught a thing. Some of the other guys began grumbling about me, too. I felt like a bait shrimp, hooked through the spine and flung out into the marsh, surrounded by predators.

The sun was at 2 o'clock, about time to call it a day. I made another cast into the brackish water and *Toink!* The line went taut.

I got myself a honey hole.

A honey hole is a spot in the water teeming with fish; sometimes you can even see them swarming if you're close enough. The trick is not to look, because you'll see the fish going in open-mouthed for your bait and want to yank your line to set the hook. Trust me, it won't work. You have to do it all by feel: put a finger or two on the line right above your reel and wait for the microscopic tug.

I kept casting into the honey hole, bringing them in one after the other. At first I was afraid it would make E.J. and everyone else even more mad, but it had the opposite effect. Though they had caught nothing themselves, the men stopped grumbling.

"Shit, girl, pull your rod back, set the hook—thassit!"

"Hoo-wee, dis goil can *fish*."

As the men took turns removing fish from and rebaiting my hook, the tension evanesced into the skeeter-filled air. E. J. Waguespack sort of apologized for doubting me.

"C'mere, dawlin'. Git ya Uncle E.J. 'nother beer."

Single-handedly, I ended up catching over twenty fish that day, mostly redfish. The black, eye-like dots on their tails provided an optical illusion that virtually doubled my catch.

"'f I had a daughter like that," E.J. said to Dad, "I'd never let her outta my boat."

Dad slipped his arm across my shoulders and drew me close, his lips almost brushing across the top of my head. I felt more alive than I ever had.

That evening Dad and I were driving home, giddy from sun and teasing each other about my already legendary haul.

"You didn't catch any-thing, I caught it a-all," I sang, sticking my tongue out at him.

The next thing I knew, my head hit the passenger-side window from the force of his slap.

"Don't you EVER stick out your tongue at me, girl," Dad said.

* * *

"Bert Laschke called collect," Mom said when we walked in the door. "He's flying into the New Orleans airport tomorrow at five and wants you to pick him up."

Bert Laschke was a friend of my dad's from Wisconsin. One year when the construction business was slow in Kansas City, Dad took some work up there. Becky and I went up to visit him once, flying Braniff Airlines with Grandma Vivian, my dad's gambling, scotch-drinking mother.

This was post–Pucci era Braniff. No traces lingered of its former splashy, flashy, jet-setting Euro-glory. The cabin upholstery was a harvest-gold-and-rust-colored print of teakettles and rotary eggbeaters. The stewardesses wore dolman-sleeved polyester charmeuse blouses and mustard-colored slacks stretched high and tight over their pooched-out tummies.

I didn't know what I was missing. Dull earth tones in clothing and décor were all I'd ever known. I was just thrilled to be going anywhere, especially when it involved driving through a cloud.

Our father was living in a trailer outside Milwaukee, subsisting on a diet of Triscuits, butter, whiskey, and Wink soda. Grandma fell in step right alongside. Becky and I spent the week eating cold hot dogs and watching soap operas. I, for one, approved of this new snacking lifestyle.

On our last night in Milwaukee, after a fine dinner of Cheetos and root beer, Dad took us to a roadside carnival. There was a cage up on wheels, like the car of a circus train, with an animatronic gorilla in it. When you came up close, he would roar and swipe at you. I wouldn't go near it.

"Ha ha, look at this!" Becky said, reaching through the bars and pulling the dirty, matted fur on its arm. "Come on, Sarah. It's fake. Don't be such a fraidy-cat."

The gorilla swiveled on its base and bared its plastic fangs. If Becky hadn't jumped away, her arm would have been stripped down to the bone.

"Duh, I know it's fake," I said. "It's just that I'd rather get a balloon."

Down the carnival midway, a swaying clown was making balloon animals. I waited ten minutes as he wrestled with a red tube.

"Here ya go," he said. "A pretty lil' bow for a pretty lil' . . . girl?"

It looked more like a knot than a bow.

"I have an idea," I said to Dad. "Let's walk around the other way to get back to the car."

I didn't want to get within fifty feet of that gorilla again.

"We're going back the way we came," Dad said.

* * *

When Dad was up in Wisconsin, my parents were in the midst of a trial separation. They kept it unofficial, not telling us kids. The fact that Dad hadn't come home for Debbie's birth was viewed not as a slight, but as a relief. His coming home was generally viewed as a bad thing, because we had to clean up the house.

Once we moved down south and my parents decided to stay married a bit longer, Dad didn't go to Wisconsin anymore. There were plenty of local opportunities for work. He kept in touch with Bert Laschke, singing the praises of Louisiana's booming construction business. Bert Laschke liked what he heard.

Bert lived with us for three months, sleeping in Becky's canopy bed. She moved in with me, forcing me to tickle her back for half an hour every night.

Bert was diabetic and always carried a piece of candy in the chest pocket of his Munsingwear shirts.

"Can I have that?" I'd ask, reaching into his pocket seventy-nine times a day.

"Nooooo, need to hold onto that," Bert would answer in his soft hoarse voice, sipping diet Shasta Grape. "Might save my life one of these days."

Bert didn't say much else. Every once in a while, he'd come up to Mom and ask, "That pimple behind my ear ready to pop?"

If it was ready, she'd dip a needle in rubbing alcohol and pop it, but I could tell she didn't like doing it.

After Bert moved in, I didn't go fishing as much. Once, he and Dad and I went out on the boat, but it wasn't the same. I caught nothing. There were only Tab and Fresca to drink. On the drive home, I had to cram myself into the sliver of a backseat in Dad's truck.

"Hey Laschke, what do you call a Polack martini?" Dad said.

Bert grunted.

"A beer with a booger in it!" Dad laughed.

Bert snorted, unfazed. By now he was used to my dad's stabs at his heritage.

"Damn, that Jaclyn Smith's a knockout," Dad said.

"Huh?" Bert mumbled.

"You know, that gal on *Charlie's Angels*," Dad said. "I don't like that Farrah. She's got a jaw like a man. Square."

I had nothing to add to this topic. I preferred Kate Jackson, who played the throaty, no-nonsense Angel named Bree.

"And that one on *Dallas*—Sue Ellen?" Dad continued. "Her eyes're too far apart, like a space alien."

"Hunnnnn," Bert said, ever the sparkling conversationalist.

"Seerah," Dad said, speaking in a Wisconsin accent like his Best Friend Forever Bert Laschke. "Who was that little gal you were friends with down in Houma?"

"Chantelle Benoit?" I said. "She wasn't really my fr—"

"Nah, not her—that redheaded gal that you drew all those dirty pictures with," he grinned, practically turning around to face me as he nudged Bert, who let out a soft snore. "Ol' Megan What's-her-name, a real firecracker."

I felt my head and neck turn hot.

Dad gave Bert a hard poke and laughed when he startled awake.

"Seerah and this little gal drew some filthy pictures together, I tell you what," Dad said. "'Member that, Seerah?"

I felt like there was a rock in the back of my throat.

"I said, REMEMBER THAT?" Dad grinned back at me again.

Bert shifted in the passenger seat.

"No," I whispered.

I closed my eyes and pretended to be asleep the rest of the way home.

* * *

During the last month of Bert's stay, his wife Vicky came down to take a look at Louisiana for herself. They bought a pickup truck with a camper top on the back. They slept in Becky's twin canopy bed together. She shaved his back out on the patio. Vicky hand-washed her negligees in our bathroom and hung them over the shower rod to dry.

Vicky commandeered the kitchen, whipping up saccharin-infused delicacies for her husband. She used a butter knife to scrape the gunk out of the seams in the top of our dining table, tsking and showing the gunk to my mother.

"See this? Ooh, this would just drive Bert crazy!"

Mom loathed Vicky Laschke.

"Listen here, toots," Vicky would say as they washed dishes after dinner. "Bert needs it every night, and I'm gonna be the one to give it to him. That's all there is to it. I'm the kind of woman who believes in satisfying her man. If you're not in the mood, for chrissakes, take a look at a *Playboy*. You know, your hubby was a real swinger up in Milwaukee!"

One night my parents had a loud fight back in their bedroom. The next morning, hanging over their shower rod was a white nylon nightgown with a see-though panel at the bust.

Soon it was official: Bert and Vicky had decided to stay in Louisiana.

"We're buying that lot over on Piney River Drive," Vicky said, unfurling the blueprints for their custom-built home on top of our filthy dining table.

"Great," Mom said. "When are you leaving? I mean, when are you leaving our house?"

Vicky looked at my mother, not sure whether she intended her remark to be nasty. Mom smiled, jutting out her chin and stretching her neck to the side in a friendly, inquisitive way.

"We rented a trailer. We're going to pick it up and bring it over to the property later today. Now look," Vicky said, tapping the blueprints. "We'll have three baths! I don't see how you can live with just two."

The Laschkes packed up their things, negligees and all, and brought them out to their camper truck.

"Wait just a cotton-pickin' minute," Vicky said, examining something on the passenger side of the truck. "What's this?"

Dad ran over to her.

"Bert!" Vicky wailed. "Someone scratched it with a key or something!"

On the side of the dark green Ford pickup, over an area about the size of a piece of loose-leaf paper, were a series of light green zigzags etched furiously into the paint.

Dad turned around, opening and closing his hands in helpless embarrassment.

"You did this!" he said, picking me up by the collar of my shirt and shaking me.

He grabbed me and dragged me into the house, down the hall, to my bedroom. He smacked me down to the floor and slammed back out to the driveway to help the Laschkes finish loading up their truck.

I lay there on the floor, throughout the afternoon, listening to the sounds of a busy household slowly resume: doors closing with increasing gentleness, pots and pans clanking against the stovetop, the "tick-tick-tick" at the end of *60 Minutes* fading into the theme song from *Welcome Back, Kotter,* the rustle and shuffle of my father's bag of Corn Nuts, and far away—perhaps in the woods behind the house—my sister Hannah yelling, "Stop pulling my hair!"

I lay there, in the spot where I fell when he hit me, into the evening, until the last teeth were brushed and the bathroom faucet across the hall dripped its last drop, still wondering:

Did I do that?

What Would Mr. Goodbar Do?

On the first page of my baby book, right under my weight (7 lb., 9 oz.) and my length (21 inches), below the blank space where my photo was supposed to be glued, Mom wrote:

I say "Jee-Jee" fondly when looking at pictures of the Baby Jesus.

According to her, it first happened when I was three months old. It's possible Mom had spent hours flicking a stack of prayer pamphlets like flashcards in front of my unfocused baby eyes, hoping religious precocity would give me a leg up on salvation. It's far more likely she parked my playpen under the two-foot-by-three-foot portrait of Jesus that hung over our fireplace. This Jesus made an ideal babysitter: he had deep-set, kindly eyes that seemed to watch you no matter where you were in the room, and he didn't charge a thing. One hand was held up, palm out, like Tonto saying "How, Kemosabe." His other hand gracefully supported his own exposed, bloody, thorn-riddled heart.

Oh, that Jee-Jee.

The next entry in my baby book is also written in first person from my perspective:

Today I fell out of my crib on my head AGAIN! Boy did I ever get a big purple goose egg right in the middle of my forehead! But I just smiled my gummy smile! From now on, I hope Mommy remembers to put the side of my crib up so this doesn't keep happening!

Those repeated blows to my head must have jumbled up my early religious training. By the time I started kindergarten, I believed that if you pestered God enough for something, no matter how mundane, he would have to give it to you, if only to preserve his good name.

Please God, let this teddy bear speak to me, I prayed every night for three solid years. And I didn't even like this particular teddy bear. He was a stiff, scratchy thing that let out a cloying moan when placed in the supine position.

God, I swear I won't tell anyone. Just make him say something.

Nothing.

How about "Hey there, sweetheart."

Still nothing.

Okay, how about a smile? A wink?

Never happened.

Jesus, God! Come ON.

Apparently God didn't play the Dare portion of Truth or Dare.

Chicken.

✵ ✵ ✵

Sister Giles taught third-grade at St. Aldric's. She was very old and wore über-traditional nun garb: the long black multilayered dress with a pleated white wimple around her face and a black veil that reached her ankles, one of which was twisted and gnarled from some unspeakable childhood accident. Despite her severe stutter and chronic, rattling cough, she always emcee'd the annual 4-H Club talent show, microphone flush against her lips:

"Th-th-th-that was Janie McCrory tap dancing—*horrr horr horr horrr*—to 'N-n-n-night Fever' by the Bee Gees. Next up is Sarah—*horr horrrr*—Thyre with an improvised gymnast—*horrr*—tic tap-jazz-ballet routine to 'Give My Regards to Broadway,' accompanied on piano by her sister B-b-b-becky."

Though technically she hadn't entered the competition, Becky was awarded first place. At first I wanted to demand a recount. Then, I reasoned that what I was doing was so avant-garde, a trophy would have negated it.

When discipline was called for, Sister Giles bent kids over her walker and kicked them lightly with her twisted leg, her words softly trickling through her tricolored moustache, "This s-s-s-s-spanking won't hurt your poddy, just your p-p-p-pride."

The fourth-grade teacher Sister Bartholomew had a paddle named Janet II; Janet I had snapped from overuse. Usually, all she had to do to restore order was feint toward the drawer where Janet II was kept. If an actual whipping was necessary, Sister Bart, using Janet II as a prod, would herd the offender into a supply closet at the rear of the classroom. The whacks and screams coming from that closet kept Janet II in her drawer for weeks afterward. Whether I or II, Janet was the least formidable weapon of Sister Bart's arsenal.

In fourth grade, Dewey Meeker was the ultimate class nerd, the geek *de résistance*. He was a mole-speckled runt with long, delicately tapered fingernails, which he used to pick his nose and line up the findings along the edge of his desk, keeping count with pencil hash marks that he erased at the end of each day. Everyone teased him, even Sister Bart.

One day during Reading Comprehension, he raised his hand and asked, "Sister, can I please be excused?"

"I don't know, CAN you?"

"Could I—" Dewey began again.

"Don't you mean '*May* I,' Mr. Meeker?"

"I need to be excused, please," Dewey said, his head swiveling around as though he were expecting a blow from behind. I recognized that move.

"You've got tell me WHY you need to be excused, Mr. Meeker. If it's a trip to the bathroom, you know that's against my policy," Sister Bart smiled. "What would happen if we all went running out to take our little breaks whenever we felt like it? Well, Mr. Meeker?"

"But Sister," Dewey began. It was too late. Urine pooled in his seat, soaking his khaki uniform trousers. He had to sop it up with paper torn from his Reading notebook.

I felt sorry for Dewey, but I couldn't openly extend any sympathy. I wasn't exactly popular. I couldn't risk associating with the class Booger Bag.

A few months later, Sister Bart announced we would be doing a special exercise. Each of us was to write down the names of three people in the class we admired and why we admired them. When we were done, we would turn them in to Sister Bart. She would then read them aloud without identifying the author, to protect our anonymity. It was the crude dawn of self-esteem awareness in education.

"I like Tracy because she's cute and sweet," Sister Bart read, rolling her eyes. "I like Keith because he's cute and nice."

She sped through a few more thematically similar pages. She picked up a page with a smooth, clean edge. Though torn from a 29-cent notebook, *someone* had thoughtfully bothered to trim the ragged edge of this page.

"I like J.R. because he wears an eye patch," Sister Bart said.

It was mine.

"I like Corinne because she prays for two hours every day and I sure hope God notices because she could use some help with that thing growing behind her ear, poor thing."

Everyone started looking around the classroom, snickering. Sister Bart paused, looking heavenward for a few seconds.

"I like Dewey," she continued, now in a fake little girl's voice, "because everyone makes fun of him for picking his nose, and he doesn't care."

By now everyone was laughing out loud. Mickey Foote, who sat directly in front of me, turned around and sneered, "I know you wrote that, ugly."

Duh. Sister Bartholomew stared right at me while she was reading it. Everyone knew. Dewey Meeker never spoke to me again, which was just as well. Fuckin' Booger Bag.

Faith and Hope were still okay. I was through with Charity.

✲ ✲ ✲

On the first day of fifth grade, I approached the school bulletin board in fear for my life. I was about to find out who would be my homeroom teacher: Mrs. Faulk or Sister Titus. I had prayed all summer to get Mrs. Faulk, the brusque chain-smoker with a sibilant *s*. She directed the Christmas concert every year.

"Heeeeeeem," she'd blow into her pitch pipe at the start of each rehearsal of Handel's *Messiah*. "Remember, studentsss: I want your alleluia'sss to be very crisssssssp."

The walls of Mrs. Faulk's classroom were plastered with posters of adorable puppies and kittens saying, "Hang in there" or wondering how they would ever fly with the eagles when they were stuck with a bunch of turkeys. Beneath each, she pasted cute hand-lettered signs inscribed with the names of her male students.

According to school legend, Mrs. Faulk had hated girls ever since the year she'd had her own son Portis in her class. Portis ran around snapping the bra straps of girls who developed too early. Some of the bustier girls reported Portis's trespasses to Principal Lusco. Dressing down the tattlers in front of the whole class, Mrs. Faulk delivered a speech so memorable, any St. Aldric's pupil worth his or her salt could recite it as perfectly as the Lord's Prayer or the Pledge of Allegiance:

> "I feel as though you have stabbed me in the back, twisted the knife, and poured salt into the wound."

If I was praying to get Mrs. Faulk, Sister Titus must have been pretty bad. She seemed to hate girls *and* boys. She was short and round, with cat's-eye glasses, a long, flat nose like a proboscis monkey's, and football-shaped breasts that swung in opposite directions when she ran. In her calmer moments, Sister Titus spoke with a genteel New Orleans accent, pronouncing Vaseline "VAZZ-uh-lun" and gasoline "GAZZ-uh-lun." However, should the smallest slight tap into her wellspring of pure, distilled rage, Sister Titus shot off like a geyser.

If a boy forgot to wear a belt with his uniform, she made him stand before the class while she roughly threaded a pink ribbon through his belt-loops and said, "How you like that, little *girl*?"

Once, out on the playground, my sister Hannah's shirttail had come untucked. At St. Aldric's, this bordered on mortal sin. Sister Titus rammed it back into Hannah's skirt, telling her, "You look like a whore on Bourbon Street."

Boy or girl, Sister Titus would blame her mood swings on you because "you and your whole family make me sick." I heard that once, she had thrown a kid down the stairs. In his desk. He was the kid who usually got a free pass because he had the BB pellet permanently embedded in his eyeball. Sister Titus spared no one.

Please God, please. I could be the one that changes her mind about girls. Please let me get Mrs. Faulk.

I got Sister Titus.

Almost immediately, all around me, children fell in puddles of tears and blood. I had to cook up a survival scheme, fast.

"Mom," I said, clutching my gut, "I have diarrhea."

"Again? Ech, you probably caught it from Grandma. From now on, Hannah's putting the sulfa cream on Grandma's sores, not you. Let's go see Dr. Camp," Mom said, her eyes flickering with Munchausen's syndrome by proxy.

Grandma was our grey cat. She was always getting into fights with other cats in the neighborhood. Strips of skin hung down over her eyes, and the wounds would suppurate for weeks. I enjoyed putting the cream on, because it gave me a daily opportunity to play Animal Hospital Show. I never just played school or played house; I played School Show or House Show. Everything was televised.

Dr. Camp and my mother decided that my diarrhea was a full-blown case of mange-induced nervous stomach. I brought a note to Sister Titus that said I was to be allowed to go to the bathroom whenever I wanted.

"Sarah, petite cher," she said, pulling me close into a cloud of bacon-y breath. "Listen up, you don't even have to ask, just get up and go when you feel like it."

Thus, I circumvented the typical nun policy of no bathroom breaks. I thought my Get Out of Class Free card would make me look cool to my classmates, like the girl with the mystery ailment who got to chew Aspergum whenever she wanted.

I enjoyed a blissful week of getting to leave class whenever I wished, until Mickey Foote pushed me up against a wall and asked, "How come you get to do that?"

"I have permission," I said, omitting the specifics.

By the end of the day, everyone was calling me the Poop Freak.

I volunteered to stay in during recess and wash the blackboard. When kids made fun of me for trying to be the teacher's pet, I deflected their taunts with the most accurate, non-flattering imitation of Sister Titus I could devise.

"Marie Louise Lapeyrouse, why you tardy again? I'll tell you why. Because you and your whole family are so damn lazy, that's why. I seen your daddy, he can't even run a comb through his hair before he leaves the house," I'd say, saliva foaming in the corners of my mouth. "WHEW. Class, I'm sorry Marie Louise made me yell like that." Pause. Sweet smile. "I just can't stand the way she comes in here with her wrinkled blouse. You can tell her mama don't love her enough to iron it."

I'd lick my lips, reach my hand up under my blouse and scratch my boobs vigorously: up over the shoulders where the bra straps dug in and down and around and around the nipples, and finally under the flaps in a fanning motion. This is something Sister Titus did when we were alone in the classroom together and she thought no one was looking. Even though I was washing the blackboard, I was *always* looking. I watched Sister Titus the way I kept an eye on that teddy bear I was always asking God to make talk. Lately I felt uneasy around Teddy, as if I'd pestered God too much about it, and He was going to let the Devil possess Teddy, just to teach me a lesson.

After a few standing-room-only performances of *Sister Titus—Alive and Spitting,* I finally convinced my classmates that washing the blackboards and watering the plants and bringing in homemade spritz cookies on Fridays was all research. I had to get intimate with my subject in order to deliver the most believable portrayal possible. I was doing it for their enjoyment and for the sake of my art.

I left fifth grade with a vocational grudge. Sister Titus was such a bad nun, she made me want to become a nun just to show her how it was done. I entered sixth grade at Saint Aldric's on a burst of piety. I befriended Corinne Crowley, the class goody-goody, the holy one with the growth behind her ear. She had a long, mournful face that just begged to be surrounded by a wimple. We shut ourselves up for hours in her bedroom, dissecting year-old Easter eggs and making crucifix necklaces out of paper clips and gum wrappers, and fantasizing about becoming nuns together. We couldn't agree on what order to join.

"When I'm a nun, I'm going to live in a cabin alone in the woods and do nothing but pray all day long," Corinne said.

"Oh, not me, I need to get out! I definitely won't be a Carmelite, because they have to wear that poo brown color all the time. Is there an order that wears purple or lavender?" I said, taking off my glasses and looking up close at myself in Corinne's vanity mirror. "Which kind of nuns are allowed to wear contact lenses?"

"Who do you love most, Sarah?" Corinne asked, with palpable yearning. "The Father, the Son, or the Holy Spirit?"

"Definitely the Son," I giggled, picturing Jesus as Alan O'Day, the bearded, leather fedora–wearing deejay who sang the pop hit "Undercover Angel."

"Oh, yes!" she agreed.

Corinne deferred to me on such catechismal matters because I had received the sacrament of Confirmation back in first grade, at St. Lucy's in Kansas City. In Louisiana, you didn't get confirmed until eighth grade, sometimes later. Confirmation was receiving the Holy Spirit via some oil rubbed on your forehead by an authorized priest, or if you were lucky (like me), a real live bishop. My mother had sewn my confirmation dress herself from some red and white calico and trimmed it with lace. The dress buttoned at the back of the neck with three tiny flower-shaped red plastic buttons.

I hated those buttons. I hated ALL buttons. I much preferred the efficient, metallic zing of a zipper. Buttons felt slick and slimy to me, and smelled like expired cottage cheese. Molding them into a cute, flowery shape couldn't stop them from making my skin crawl.

By the time it was my turn to be anointed by the bishop, I had pulled my collar around and bitten the buttons off. I held them under my tongue throughout the remainder of the ceremony.

"Do you reject Satan and all his works and all his empty promises?" the bishop asked.

"I do," I mumbled, trying to keep the buttons from clacking against my teeth.

When Corinne and I found a dead pigeon on the school playground, we danced around it unselfconsciously, singing, "Rise up to Heaven, Hallelujah!" For about thirty minutes. Let the other kids point and laugh. What did we care? They were all going to Hell while we were on the bridal path to Jesus Christ.

On sleepovers, Corinne and I would hold hands and stare into each other's eyes while singing along to the latest Reverend Carey Landry album:

Hi God, how d'ya feel today? Can you hear me, God?

The song went on to list all the things we had to be thankful for: salads, hamburgers, ice cream, our lives.

Sister Lazarus was the nun who turned us on to Reverend Landry's music. She was the sixth-grade teacher that fall. Sister Lazarus was thin, with a parted iron gray Afro that showered dandruff onto the top of her thick, black, Carol Channing–style glasses. This made her appear gentle and placid as a snow globe.

Reverend Landry's music was hip compared to "Be Not Afraid" and "Now Thank We All Our God." Sister Lazarus tried her best to jazz up the musty religion curriculum, which was usually just Ten Commandments this, Transubstantiation that, and droning old hymns best sung in an old lady's high, tuneless monotone.

That year at the Christmas concert mass, instead of the "Hallelujah Chorus," we sang "Day by Day" from *Godspell*. We baked our own hearty Eucharist with stone-ground wheat and raw honey and decorated the altar with abstract expressionist murals. At last, the Age of Aquarius had arrived. Better ten years late than never.

I left for Christmas break high on this wondrous new hippie love religion. When I returned to school in January, most of the nuns had disappeared, called to some vague woodland location for an open-ended "retreat." While the nuns were out for reeducation, we would be taught by a series of sensuous-sounding "laypeople."

The entire class enjoyed sinking our fangs into the first few of these greenhorns. Then came Margie Reed. She swanned into the room wearing a maroon skirt so crisply knife-pleated it looked like a lampshade. Perching on the edge of Kevin Schwab's desk, she trained her sparkling Christian eyes upon us and smiled.

"Ah, the Book of Genesis," she sighed breathily. "You know, they say God created the world in seven days . . . but a day to us might be *a thousand years* for God."

A murmur rose up among us students. Was she saying that the Bible wasn't *true*? The mere thought freed our minds like a long, deep bong hit. Was she saying it took God 6,000 years to create the universe and 1,000 years to rest? Blasphemy! Heresy? Whatever. Hooray!

The next day, I was in my seat way before the second bell, the one that meant you better hustle if you didn't want to be tardy. Margie Reed swept into the room, taking delicate steps in her size 5 peep-toe pumps. She was wearing another lampshade skirt, this one a rich brown.

"That's a lovely skirt you're wearing," I said. "Very autumnal."

"Why, thank you," she said, smiling into space as the rest of the students files in.

From the moment the second bell rang, Margie Reed was off and running. She didn't disappoint.

"So then, Vatican II," Mrs. Reed began in her deliciously insinuating tone. "The pope wants to update the faith, right? So he calls in who? Some married couples? Noooooo. Just a bunch of priests. Priests making decisions about birth control? Tell me how that's a good idea!"

None of us could, but we were seduced by her provocative notions. I for one was such a convert, I even took it upon myself to reach out to the community and proselytize.

"Don't believe a Goddamn thing you read in the Bible!" I instructed my little sisters.

"Okay," Hannah and Debbie mewed compliantly.

Margie only substitute-taught for three days, but it was long enough for me to make a good impression. I started babysitting the Reed children. I'd been babysitting outside the home since I was eight years old. Mrs. Reed's husband Steve found my experience impressive and my rates reasonable.

The Reeds had two kids, Gwen and Katie. Gwen, a couple of years younger than I, was known at school as "Rocket Head." She had a large forehead. Technically, Gwen wasn't a real waterhead like Gerald McQueen, but it would have been outright cruelty to call Gerald "Rocket Head," and it was simply too good a name to go unused. Gwen thought she was old enough to watch Katie herself and resented me. She spent most of her time up in her room reading, expanding her skull even further.

Three-year-old Katie was pretty cute, and easy. All she wanted to do was sit in front of the record player, listening to the same alphabet song over and over. It was one of those floppy plastic 45s that came in a magazine, free. I think it was put out by the state department of agriculture. A was for apple, B was for beef, W was for watermelon, and every verse ended with: "Ask your mommy to buy some!"

Katie sat for hours listening to that record, rocking back and forth, mesmerized by what was basically a glorified grocery list. She paid no attention to the rustling and crunching noises I made out in the kitchen with the real groceries.

The only downside of the job came at bedtime, when I had to apply Katie's medicine. She had some sort of rash that never seemed to go away.

"Put the cream on my veegeeeeeeeee," she'd say, lying on her back on the bed with her legs spread.

I'd politely dab it on, looking at the ceiling and wondering if it was worth 50 cents an hour.

I had to wonder: if Margie Reed, with her marvelous ideas and magical skirts that defied humidity, had kids that were cursed with such maladies, what could *I* possibly do to escape cancer? My sister Debbie had just had a lump removed from her neck. The doctor said it was benign and not catching, but I wasn't drinking after her anymore, no sir. That girl was contagious, period.

Last year, Debbie brought home a hardy strain of lice from kindergarten. Mom shaved off Debbie's butt-length hair, and boiled our sheets and pillowcases.

"This oughta kill those little shits," she said, pouring lighter fluid into the washing machine.

At last the lice were vanquished, or so we thought. Becky and Hannah and I went to Robert's School of Beauty for $2 haircuts.

"Okay, just sit down in this chair, Sarah. My name's Nina and I'm gonna comb you out," Nina said, sinking her black Ace comb into the rat's nest on the back of my head.

Comb, comb, comb. Pause.

"Oh-oh. Omigawd," Nina gasped, scraping the comb in a sawing motion across the tender flesh of my scalp. "Sheila, get over here, you have *got* to see this. Bring Gavin, he's always wanted to know what nits look like. This is the worst case of lice I've ever seen. The other girls have 'em, too?! Jesus, call the School Board. Ewww look, they are crawlin'!"

Since then, I expected plagues and other mishaps to befall my family. But Mrs. Reed's? I couldn't shake the feeling that the Reeds must have done something awful to make God visit such physical curses upon their offspring. After Katie and Rocket Head went to bed, I dug through the drawers and cabinets, looking for a clue.

At last, in the nightstand, I found it. It looked like a tube of prescription toothpaste. Upon closer reading of the label, I discovered it was something called "contraceptive jelly." I unscrewed the tube's cap, smeared a little onto the back of my hand, and gave it a sniff. Seemed like ordinary Vaseline. As I stuck out the tip of my tongue for a taste, it dawned on me.

"Birth control's illegal in the eyes of the Church. The Church frowns on prophylactics," Mom said, giving Catholicism a giant, disapproving face.

No wonder those kids were afflicted.

God, I swear I will never use these dirty sex creams. Well, not in real life. I didn't see the harm in using *pretend* creams during make-out sessions with my pillow.

"Goddammit, Alan O'Day!" I moaned, sotto voce, in my bed at night. "You get over here this minute and put some sex jelly on my veegee, oh *yeah*."

Mystery solved, I could spend my downtime relaxing in front of the television. The Reeds had cable. *Cable* TV. At our house, we were still twisting pieces of tinfoil around the tips of a rabbit-eared antenna. Dad was too cheap to get cable and Mom was anti-television because cathode ray tubes shot radiation into your eyes. It was a rare instance in which their views dovetailed.

Once I figured out how to operate their newfangled remote control, I realized the Reeds even had that most sinful of channels: HBO! Of course they would.

Back then, HBO's offerings were rather scant, usually just low-budget slasher films. Then, one fateful night, came *Looking for Mr. Goodbar.*

I settled in to watch it, a bowl of popcorn, bag of Bar-B-Que Fritos, and can of deviled ham on my lap.

Richard Gere runs around in his underwear—okay, pretty good. But wait—then he stabs Diane Keaton repeatedly in the chest with what looked like a butter knife. Yes, they were having premarital sex and would definitely be going to Hell, but was that really necessary? Maybe Richard Gere figured he was going to Hell anyway, so he might as well kill someone.

By the time the movie was over, I had a sick knot in my lower abdomen. A nauseating, tingly knot. Was I about to go upstairs and murder Gwen and Katie? No. This was either my sexual awakening, or I was nervous that Mr. and Mrs. Reed, however slathered with sin-jelly they may be, were going to walk in at any minute and catch me watching a murder-sex scenario. When I got home that night I said three rosaries and promised never to watch HBO again.

Just to be on the safe side, I decided to stop babysitting for the Reeds altogether. I had other prospects.

Hans and Lotte Blenreich were a young German couple in our neighborhood. They ran a perpetual rummage sale out of their living room. The Blenreichs sat in folding chairs at a card table, the only furniture in the house besides the entertainment center and a double-sized mattress in the living room. They sat there all day, putting price tags on everything they owned. Sometimes their friend Rusty would be there, too, staring into space.

My sisters and I dropped by every afternoon to browse. Over the course of their sale, I amassed a lovely assortment of novelty shot glasses.

The star of my collection was an especially clever one with a boob on it that said, "Daddy's Milk." You could drink through the nipple!

Their one-year-old son Fritz was always propped on a cushion in the corner while their dog—a dachshund—licked yogurt and Ry-Krisp crumbs off his face. Soon I picked up a regular Saturday night gig watching Fritz. Fritz was allergic to all kinds of baby formula, so Lotte had raised him on goat's milk. He weighed 65 pounds and looked like a mini-Michelin Man, with roll after roll of rich creamery lard cascading down his thighs. Babysitting for him would be a workout but worth it: the Blenreichs had a Betamax.

Saturday nights, I'd roll Fritz onto the bed and fire up the Betamax. It was the first one in the neighborhood, the size of a large microwave oven, top-loading, and studded with thick, silver, pleasing-to-the-touch buttons. These were the kind of buttons I liked: the kind you could push.

As far as I could see, the Blenreichs had only one tape, *Breaking Away*. I watched it over and over and over again. The Christian in me tried to like the blond guy who sang opera and shaved his legs and raced an eighteen-wheeler on his bicycle. Secretly, I lusted after the angry, shirtless Dennis Quaid, who drove around in a muscle car and had no future. In reality, I would probably end up with the short, greasy, long-haired guy who got his girlfriend pregnant. After all, his name was Moocher.

Around the fifth or sixth time I babysat for them, Lotte made an announcement.

"Sarah, we got cable TV. It costs a lot of money, so we'll be paying you 35 cents an hour instead of 50 cents. But wait, that's not all! At the end of the night, I'll throw in two shot glasses from the sale table."

I wanted to shout, *Lotte Blenreich, I can't watch your filthy cable! Do you want me to go to Hell, along with your fat, unbaptized baby?* Somehow, Mom had figured out the Blenreichs didn't go to church. One night when I was babysitting, she came over and sprinkled holy water on Fritz, but I wasn't sure how legitimate that was.

I fought off the temptation of cable TV by digging around in the kitchen cupboards. Going on a snack hunt in this house was a joke. All the Blenreichs ever had to eat was thick, black, whole grain bread studded with currants and seeds, some cheese that smelled like belly button lint, and ten kinds of mustard in the fridge. Once I had to resort to scarfing down a whole tube of Braunschweiger, scooping it into my mouth with my fingers. I made the best of it, pretending to be Kunta Kinte in *Roots*, eating that bowl of grits.

Maybe they hid the good stuff somewhere else. High up in a cabinet in the den, behind a Make-Your-Own-Potholder kit, I didn't find anything to quiet my growling belly. I found something to stanch a different kind of hunger: more tapes for the Betamax. There were three of them, with barely legible titles scribbled on their rough cardboard sleeves: *School Bus*, *Summer Camp*, and *Sex Boat*. There was a smiley face sticker on the *Sex Boat* tape, which charmed me.

I popped *Sex Boat* into the Betamax. Huh, no opening credits, interesting. They just jumped right into the story. Two men and a woman were lounging in a hot tub, cruisin' along on the Sex Boat. Sure they were naked, but I was playing it cool, acting like it was no big deal. I had the volume turned down all the way. Fritz snored softly on the bed next to me.

All of a sudden, apropos of nothing, the naked lady starts lapping at one guy's penis, while the other guy licks her vagina. I rushed the TV to stop the tape but instead found my hand reaching out to turn the volume up just a leeeeeetle bit.

"She ready, Speed?" said the guy getting lapped at.

"Her pussy tellin' me she ready, Ross," the licking guy answered.

And so on.

Yes, in the dim lights of my parents' closet, I had read in Dad's dirty paperback about Phyllis arching her back and riding the waves of Sylvia's expert touch. Yes, I'd considered reenacting the biracial "The Pearl Necklace" pictorial from the moldy porn Bitsy and I found in the woods. Still, I was innocent, sexless. I'd never heard the terms "blow job" or "eating someone out." The only thing being eaten out in my life was a tube of liver sausage. Furtively.

Mr. Goodbar had given me a tummy ache. *Sex Boat* gave me the dry heaves. I took some deep breaths—okay, one deep breath—until I felt well enough to watch a little more.

Speed and Ross were cramming their penises into the poor woman's every hole.

I was in shock. The actors' poor diction and low-class vernacular were inexcusable. Until I saw *Sex Boat,* I assumed that movie acting required a British accent and the enunciation of Ethel Merman.

"Morons," I sniffed.

I took a smug pause, the kind of pause that refreshes.

I yanked out the tape and said an Act of Contrition, but I was pretty sure it was too late. Guess I was going to Hell. I turned the TV back on. Might as well watch some cable. No use in avoiding it now: HBO. Hello, old friend.

I joined a horror movie already in progress. Some crazy old coot was running around, trying to grind people up into sausage, then feed the sausage to other people, thereby fattening them up so that he could make sausage out of them, too.

All that sausage was making me hungry. I went out to the kitchen to get a snack. When I came back, the slutty girl was about to get pitchforked in the chest.

"Ha!" I said aloud, digging into a can of sardines in mustard sauce. "Serves you right, slutty."

Oh the Places You'll Go!

A Play in One Act

We open on a family of six speeding down Route 1 toward Grand Isle, Louisiana, in their Chevy Beauville van. The van is seafoam-green-and-white-striped like original AquaFresh toothpaste. Dad is hunched forward, gripping the steering wheel, his eyes glazed over in a linear, destination-oriented travel frenzy. Hannah and Debbie are wrestling on a bench seat, fighting over a bag of Buddig lunchmeat. Mom is in the rear of the vehicle with her daughter Sarah. Becky is secretly taping everyone on her new red-white-and-blue tape recorder.

MOM: 'Kay. Here's the toilet.

DAD: Oh god no—she gotta go again?

MOM: She never did go the first time. (CB radio blares loudly) C'mon. (rustling noises) Put this here before it smashes. C'mon, skootch in there.

HANNAH: Did she go poddy? I smell poddy.

MOM: Well, Sarah already shit in the bucket once. She's shittin' again.

BECKY: She is?

HANNAH: She is?

DAD: She is?

MOM: I guess . . . (sound of window being opened) SARAH! Get it in there—you're drippin' it!

BECKY: Oh gross!

DAD: God DAMN.

MOM: Don't! (pause) Debbie has to go now.

FINIS

The Hills Are a Lie

Our Chevy Beauville was no custom conversion van, tricked out super-cherry style with a mini-fridge, a Naugahyde sleeper sofa, and a flushable, cushion-seated toilet. Nay, my friend. On car trips short and long, my parents thoughtfully provided a white plastic bucket for us to go in. Considering the van's lack of tinted windows, it may have been slightly more thoughtful to pull off the interstate and stop at a rest area or a McDonald's.

The outset of our vacations always found my father hell-bent on arriving at our hotel exactly at check-in time in order to get the most bang for his buck. On the road, all normal human functions were a liability. The man would stop for nothing. Everything we ate along the way came from a cooler between the two captain's chairs up front. In his estimation, the shit bucket was a stroke of genius. Multiple bathroom stops were "a damn waste of time," Dad said, unwittingly providing a new twist on the old saw, "Haste makes waste."

Why he became so indignant when we actually used the thing for its designated purpose, I don't know. Perhaps he should have left the cooler at home and demanded we get by without food and drink. Maybe then we wouldn't have needed the bucket at all. Travel light.

Dad hadn't gotten a Christmas bonus that year, so there would be no Disney World. My sisters and mother clung to the connotations of "Grand Isle," expecting turquoise waters and misty waterfalls like that pinnacle of Disney extravagance, the Pirates of the Caribbean. I'd been to Grand Isle before, on fishing trips. I didn't have the heart to tell them about the brownish Gulf of Mexico lapping at Grand Isle's creosote-tinged beaches.

Our room at the motel was dappled with mold. The acrylic bedspreads on the bunk beds were branded with so many cigarette burns, at first I mistook them for an abstractly embossed jacquard. Beneath them lurked damp sheets and a generous sprinkling of sand. The aroma of the place seeped into my skin and hair, imbuing me with the scent of a forgotten platter of onion rings in a giant ashtray up in the attic of a thrift store. Out by the murky pool, it was bittersweet to find no sign of my soul mate, Todd. I missed him, but I was glad he didn't have to stay in this dump, not to mention watch me suffer.

Dad's old Cajun pal Herbert Hebert camped out in his RV in the parking lot. The three of us went fishing out on Grand Isle's pier. I caught a stingray.

"Whatchoo say?" Herb said to Dad, holding up the fish carefully, to avoid its lashing, barbed tail. "Stingray court bouillon tonight?"

My dad shook his head and spat.

"Hell, I ain't putting that in my mouth."

Herb slapped the fish onto the pier. He pulled a fillet knife out of a sheath on his belt and stabbed the stingray over and over again in its belly.

"That oughta do it," Herb said, flipping the fish off the pier.

I leaned out over the railing to watch the ray fall. As soon as it hit the water, eels slithered over its bloody white underside like lines of ink.

✵ ✵ ✵

Hannah, Debbie, and I spent the month of June working on the Mudhole in the ditch out by the road. The Mudhole was a fantastical orifice, quite possibly a passage to the center of the Earth. Hannah and I lowered Debbie into it up to her armpits, ratcheting her around by the wrists to widen and deepen it. The suction was so great, soon Debbie was up to her shoulders.

"Keep spinning!" Hannah and I urged, letting go of her hands.

"I can feel something down there with my feet," Debbie said, her head rotating slowly in mud up to her chin.

"Scratch at it with your toes!" I said, all excited. "Does it feel like gold bars?"

As we hosed off under the carport, Mom inspected us, sniffing and fingering us in the manner of the village crone haggling over wares in a medieval market. Expecting the same old mother who cried "worm," I tuned her out and let the water wash over my face. I splashed and rubbed like Helen Keller doing a Noxzema commercial.

"Is that a mosquito bite on your arm?" she said, pointing at me. "Well, I hope it wasn't one of those encephalitis mosquitoes. They breed in standing water, and that puddle out front's been there for weeks."

"What's encephalitis?" I asked, snapping out of deaf-mute mode.

"Just a disease that turns your brain to liquid," she said, scampering away almost joyfully.

Please Lord, let that be water dripping out of my ears.

Mom kept herself, and us, informed of all the current maladies. Worms and noxious tunnel gases were so 1976, not to mention exceedingly small potatoes. According to her, the entire environment, both indoors and out, was out to get us.

The minute we moved down South from Kansas City, my mom hated it. She stared out at the flat, humid, marshy landscape, griping about insufficient drainage, germs seeping out of the aboveground graves, and invisible mold spores hanging in the air like Spanish moss hung in the trees. Her offhanded remarks would haunt me for years.

"Go ahead and put sugar in your iced tea. It'll just make you more hot and thirsty."

"You don't want that Bugs Bunny sheet set. Sleeping on 50/50 polyester blend'll give you nightmares."

"Says here the Superdome's sinking three to six inches a year."

The fear of being smacked upside my head at any moment had already made me a tad jumpy. After that sinking comment, I began to hotfoot around, treading quickly and lightly, as though I were walking on quicksand. I looked like an ostrich.

Granted, there was some truth in her denouncements of the climate. All she needed to do was point instructively at the thick, gold living room carpet. Mushrooms grew there.

Besides, the proof was in the plagues. Becky's eyes occasionally swelled shut with hay fever, Hannah had chronic ear infections, and Debbie's palms were so moist and prone to peeling, we'd given her the nickname Shroomy (short for Shroom Hands). Whenever either of them had a flare-up, Mom pounced.

"A-*ha*! What'd I tell told you? There's mold growing under this prefab slab house. We're six feet below sea level here. We've got to get to a higher altitude!"

Going to Grand Isle was heading in exactly the wrong direction. It was mountains we needed, or more specifically, the Mountains. Mom wanted us to live inside a Currier and Ives painting set to a soundtrack of Bing Crosby's Christmas songs. Her litanies, along with mandatory annual viewings of *The Sound of Music*, convinced me that the Mountains were the answer. The Mountains would miraculously cure all our ills.

Mom's eyes glowed watching Julie Andrews twirl her way across the Alps. Her obsession always peaked around Christmastime. She polished up our bright copper kettle and attempted to knit us warm woolen mittens.

"These are too scratchy," I complained. "I need to wrap presents. Where's that roll of green swirly foil paper?"

"We don't need store-bought wrapping paper," she said. "It makes more sense to use brown paper and string to tie up our packages."

I caught her using Dad's jigsaw to carve out a miniature proscenium from the wooden backboard of an old Social Studies Fair project.

"Hey, isn't that my Cherokee Indians project?" I protested, fondly remembering the time Mom stayed up all night, weaving together twigs into a wigwam, sewing traditional costumes for the little Cherokee doll family, cutting out leather letters that spelled out "Cherokees: A Hard Life," and typing up my report, which she had also researched and written.

"You never liked the Cherokees like I did," Mom said. "Now I'll just sew up some little drapes from Hannah's old velvet Christmas skirt and we'll be ready to go."

When she was done, she called us all together and clapped her hands.

"Children, do let's put on a puppet show!" Mom said in the clipped, chipper voice of someone who had taken a vow of chastity.

I had to admit, this was much more fun than those dull old Cherokees. I cobbled together musical story lines for my motley crew of puppets: a gingerbread man, an alligator, a dog, and a furry sock with a Kewpie doll head sewn onto it. They performed all the rolls off *Little White Duck*, a children's record of Burl Ives sing-alongs about animal disease, torture, mutilation, and death.

First, the Kewpie doll sock and the Dog acted out "The Goat," a song about a beast so insatiable, its owner has to lash it to the railroad tracks. For an encore, the other two puppets performed "The Sow Took the Measles." This song was a eulogy to a dead pig, listing all the things Burl intended to render from her sweet corpse: pickles, glue, a saddle, a thimble, and a whip. In my production, the Gingerbread Man would lovingly flay the alligator, then joyfully dance atop his lifeless body.

✵ ✵ ✵

Back in the sixties, after five years of mostly chaste dating, when Dad asked Mom to marry him, she ran away and joined a convent in Illinois.

She stayed there for two days, until the Mother Superior told her, "Nancy, I don't think you're supposed to be a nun."

Take your basic *Sound of Music*, mix in a tincture of *Star Wars*, and you get what Mom claims to have heard.

"Go forth and seek your destiny. Climb ev'ry mountain!"

"Sometimes I still wonder if I missed my calling," Mom often said.

"Why *didn't* you become a nun?" I'd say, thinking it might have prevented Dad's latest rampage. I missed the larger ramifications, like never having been born.

If my mother were Fraulein Maria, and we were the von Trapp children, that made my father the Captain. Sort of. Dad didn't use a dog whistle to keep us in line like Captain von Trapp, but once he threatened to rub my nose in something I spilled on the ground. I think it was boysenberry jam.

Dad was the one, indisputable obstacle to the singing, gamboling way of life that I deserved. Once when I was sweeping up the kitchen after supper, my nightly chore, I picked up the broomstick and shimmied back and forth, breaking into a chorus of "O Me Old Bamboo, Me Old Bamboo."

"Don't miss the corner like last night," Dad said from the doorway, startling me.

He kicked the bright copper dustpan toward me. It sliced into my bare foot.

"Owwwww," I cried, dropping to the floor.

Just a flesh wound. I began to cry anyway.

"Get up and finish, or I'll give you something to cry about," Dad said, standing over me and batting my head with a rolled-up newspaper.

Most days I lived in fear for my hide, which he was always threatening to "tan." It was a gruesome choice of words, with an attendant image of him stretching my skin over a wall in a shack and meticulously brushing it with volatile chemicals. Tanning my hide would require planning and intent.

It wasn't the bemused fondness that played across Christopher Plummer's lips as he pulled on his gloves to dance with Maria, but Dad could be affectionate.

"Fee fie fo fum, I smell the blood of an Englishman! Be he live, or be he dead, I'll grind his bones to make my bread!" he growled, chasing me around the house.

I'd shriek with half-real terror, thinking about him first placing me in a giant meat grinder, then placidly kneading me into a lump of human dough as he looked out the kitchen window, perhaps watching a nest of robins. I wondered how I would taste.

I guess Dad did, too. When he inevitably caught me, he'd pin me down on the sofa and give me a "chicken lickin'." That was a slobber shower to my face and neck that left me sticky and smelling like coffee and cigarettes. It was a far cry from Captain von Trapp joshing with his eldest daughter Liesl over whether she could stay up late and taste her first champagne.

Mom got it into her head that we needed to take the Mountains like people take the waters at Baden-Baden. Her sister, my rich Aunt Carol, had invited us to visit her condo in Vail. There was plenty of room for all in its glamorous-sounding "sleeping loft."

"Can I please take the girls to Colorado for Some Mountains?" Mom asked Dad, her eyes shining with a facsimile of Austrian naïveté. "The kids need some nature to dry up their mucus."

"What, Grand Isle not good enough for you?" Dad answered, slowly twisting her wrist until the veins on it popped out even more than usual. "Hand over that checkbook."

He hid it somewhere in the house. My sisters and I split up, searching high and low, but we couldn't find it. We were left without funds for a week or so. Over meals of saltines and Jell-O and canned beets and Vienna sausages from the back of the pantry, we plotted our trek over the Alps—er, to the Rockies. Inspiration is 99 percent starvation. What wasn't killing us was making us stronger, hardy enough to survive whatever the Mountains lobbed our way.

By the time Dad gave back her car keys, Mom had cooked up a scheme to get her poor, invalid, and now malnourished children to our destination in spite of spousal adversity. It was a matter of life and death.

First off, we needed cold hard cash. Mom sent me into Rebel Savings and Loan with a forged check drawn on my father's secret bank account and my not inconsiderable acting skills.

"Excuse me, ma'am," I said to the teller, all business. "May I cash this, please?"

"Who's this name on the check, your daddy?" she said, looking at me through the rims of her reading glasses, her eyes half-magnified, half not. "Where's he at?"

"Well, you see, he broke his leg and is laid up out in the car and just sent me in here to cash that check, so can I get that one and a half thousand dollars in fifties and hundreds, please, ma'am?"

It wasn't a complete lie. My father had a broken ankle, back when he was twenty-one.

I bent my knees a little, poised to run in case she decided to look out the window for confirmation. I was always ready to be called a liar, especially when I was actually lying. I sent Mom an ESP message out in the car: *Get down. Get down.* But the teller just smiled and handed over a big wad of cash, along with a wink and a cherry lollipop.

"Look at all that money!" I said back in the car, watching Mom fan it out like a hand of gin rummy.

"Let's go buy our tickets," she said, her narrowed eyes rendering her less fraulein, more frau.

The morning we left she crafted a note to Dad, telling him dammit, we deserved Some Mountains in our life and could you please feed the fish once a day. Scooter was our goldfish, alone in his bowl with only a neon tiki hut for company.

"Put these on," Mom said over breakfast, proffering matching wrap skirts that she'd whipped up on her pedal-powered Singer sewing machine. Her right calf muscles bulged more thickly than her left.

Mauve denim had been on sale at the fabric store. At home or on the road, the wrap skirt was Mom's personally preferred garment, as it provided ample ventilation and easy hygiene.

"But I want to wear my jumpsuit!" I protested, holding out for the purple zippered one-piece I wore every day, dirty or clean.

"Oh, I'd wear the wrap skirt if I were you," Mom said, sailing back and forth in last-minute preparations for our trip. "Unless you want to end up a Ruth Rotten Crotch."

A few hours later, Mom and my sisters and I walked over to the gas station that served as the town's bus stop. Mom shook a big bottle of Shaklee vitamins at us.

"Did everyone take their brewer's yeast?"

"Yes," we lied.

Mom had recently bought into Shaklee, a sort of Avon/Amway for selling all-natural vitamins. She never had time to go out and sell them, so we had crates of vitamins stacked up in our carport. Brewer's yeast contained lots of B vitamins. Mom was always forcing it on us. You had to chew four dime-sized tablets three times a day. They tasted of corrugated cardboard with a top note of horse manure.

Our bus pulled up. Clad in our matching mauve wrap skirts, we boarded a Trailways bus to Colorado.

O, bus travel! Mom made it sound so pastoral, so rife with romantic possibility.

"Just wait 'til we get situated on that bus and watch the scenery roll by," she enthused, chewing gum with her front teeth. "There's nothing like it."

Yes, there was nothing quite like watching the scenery roll by, but watching paint dry had to be a close second. As spiritually fulfilling as travel by bus sounded, the reality of it was ever so shockingly different.

When I pressed my nose against the window to watch the scenery go by, the air vents along the bottom of the pane shot hot fumes that smelled like whiskey and cigars straight up into my nostrils. When I tried to take a nap, my cheek stuck to the seat, which appeared to be upholstered in a flocked brocade of gum and pubes. I had no choice but to remain awake, turn to the inside of the bus, and interact with the other passengers, whose commonness, I feared, would cheapen my travels.

I had imagined a bus full of women with frosty winged bangs, wearing gloves and hats and smart traveling suits with shoulder pads, amusing me with their snappy patter and sharing delicious food from their wicker picnic baskets. Instead, I got men with bangs, no wings.

The bus was a bust. If I had wanted to see shirtless poor people argue among themselves, I could have stayed at home and hung around the Winn-Dixie parking lot. Still, the trip wouldn't be a total loss if I could find someone to make out with.

Within the first hour on the bus I'd already scoped out a potential boyfriend, a sad-eyed waif named Dallas. Dallas was traveling with his dad, Mac. Mac was delivering his son to his ex-wife's house in Oklahoma City for the summer, and he took the bus 'cause he was gonna "show that bitch."

Apart from the venom he reserved for his former spouse, Mac was friendly and garrulous, the self-appointed onboard hospitality envoy. The moment we sat down he came bounding up with a jumbo bag of Funyuns and a Playmate cooler full of beer.

"Ya see this?" he told us, jangling a purple felt Crown Royal bag full of change that hung from his belt. "When we get to Amarillo, I'm gonna let these pennies fly!"

"Wooooo!" I squealed, not knowing from whence this noise came or even what it meant, but fairly sure it was the polite response.

Mac popped Mom a beer, tossed us a handful of Funyuns, and moved on down the aisle. Unfortunately, Dallas didn't take after his daddy. I soon found that outside of his ritzy name, a Ziploc bag of colored pencils, and a mesmerizingly mournful face, Dallas had zero personality.

An older, pimply teenager named Randy caught my eye. He was on his way home from reform school in Bay St. Louis, Mississippi, and had a stack of *Mad*, *Cracked*, and *Crazy* magazines. If I squinted a little, it blurred out his deep cystic acne. My nostrils pricked up at his sweet Vidalia b.o. A lacy array of blackheads, fine as filigree, lent some charm to his otherwise porcine nose. In the absence of any other prospects, he would do.

"*Mad*'s okay I guess, but I really like *Crazy!*" I said, tucking my chin and batting my eyelashes like Annette Funicello. "I mean, *Crazy*'s just . . . dreamy!"

"Take 'em all," Randy grunted. "You kin take mah seat too. Lesswitch."

He wormed past me into the aisle, slipping into the seat next to my mother. I thought he was being extremely gentlemanly, surrendering both his fine reading material and his softly scented seat.

It was all a bone being thrown to a dog. Sure, he was interested, but not in me. Randy was just using his fancy magazines to distract me so he could put the moves on my mother. My blood boiled as I looked over the top of a *Cracked* to see her laughing and patting him on the arm, a little suggestively, if you asked me.

"Oh Randy, you will *too* get your life back together!"

Mom got all the action on that bus. There were plenty of guys, young and old, hot for a thirty-nine-year-old mother of four. The odds-on favorite was Roy, a middle-aged nightclub owner from Georgia who promised Mom, "Y'all come to Atlanta, I'll set you up!"

Liking the sound of this, I encouraged their busboard romance. This time I gladly gave up the seat next to my mother.

"Mr. Roy?" I said, draping myself across his lap. "If we come to Atlanta, can I live in a mansion and have a monkey for a pet?"

"Sugar, you know Roy's the big man down there, now we just gotta convince your mama of that!"

Everything was going just great until the second day, when Mr. Roy made an unfatherly spectacle of himself. Around 4:30 AM, he woke up the whole bus, out of breath and angry at nearly being left behind at the Iron Skillet truck stop restaurant.

"I jest set myself down to have me some sausage 'n' eggs, look out t'window and the bus is leavin' me!"

He was all sweaty from running after the bus for a quarter-mile or so. He stood in the middle of the aisle, bellowing on and on about his lost 39-cent breakfast special. It was then that I noticed Mr. Roy was a bit . . . *short.* His head missed the overhead luggage rack by a good ten inches. I had mistaken his three-piece suit and Vandyke beard for dapperness. Dawn's early light streamed through the windows as he shook his Lollipop Guild fists and stamped his tiny Rumplestiltskin feet.

I decided I didn't like the idea of having him for a daddy after all. I slipped back into the seat next to Mom's.

By the time we hit Raton, New Mexico, the pickins had grown mighty slim. My mother dozed, taking a well-deserved break from being the Sexiest Woman Alive. Aside from a few stoic Navajos sneaking sips of clear liquid out of Mason jars, the bus was empty. My sisters and I decided to play Hospital Bus, wrapping up our broken limbs in casts made of the hygienic wipes my mother had brought along to complement our wrap skirts. When we ran out of those and there was nothing left to do, I dangled my foot over the edge of my seat.

"Hey, Shroomy," I said. "Come chew on my sock."

Debbie trotted right over.

"Whoa, now that's too much," Becky said as Debbie knelt in the aisle next to me.

"It's okay, I like the flavor," Debbie chirped, her mouth muffled around my white cotton Gold-Toe, black on the bottom with Trailways residue.

"Go get me that apple juice outta the cooler," I yawned, flicking the top of her head.

I accordion-folded the inside back cover of Randy's *Mad*, turning a picture of a verdant meadow into a grotesque rendering of a toxic waste dump.

When we finally got to Vail, most of our luggage was missing, except for one piece that was busted wide open. We stood bedraggled in the quaint village square next to our overflowing, duct-taped suitcase. Aunt Carol and her daughters came to pick us up.

"Well well well, look what the cat dragged in!" Aunt Carol chuckled from around her Virginia Slim. She was wearing a three-piece gold Lurex sweater suit and her hair was styled into a frosty swirl.

My cousins, Wendy and Lauren, waved shyly. They wore matching embroidered vests and huarache sandals.

If my mother was Fraulein Maria, humbly content in her shabbiness, then her sister Carol was the Baroness. She and my Uncle Leland belonged to a swanky country club, with a golf course that had green grass on it year-round. A visit to their house in Kansas City could entail playing ping-pong while listening to "Dance 10, Looks 3," the tits and ass song from *A Chorus Line*, then snacking on classy tidbits like Green Giant button mushrooms dipped in sour cream, and finally, taking a long, well-deserved soak in their Jacuzzi.

Yet Aunt Carol had come from the same home as my mother, where eleven children slept in two sets of bunk beds. Nearly every night, their father went out and got so drunk, lots of mornings found his car parked in their front yard, vomit dripping down the driver's side door. In winter, it froze.

That week in Vail, money was tight. Mom had spent most of that pile of cash on our bus tickets. We had to save some for meals on the return trip. The only activity we could afford was one twenty-minute horseback ride up a mountain trail.

Mike, the guy who ran the barn, wore a floppy leather hat and John Lennon glasses. We sized each other up.

"Any trail riding experience?" he asked.

"Oh yeah, lots," I bragged, wondering what kind of car Mike drove, and whether it had room for four kids.

"Uh-huh, right," he said, leading me to a speckled paint named Cosmo.

"What a fine day for an exhilarating ride on a high-spirited filly!" I said when Mike put the reins in my hand.

Almost right away, Cosmo wandered off the trail to eat grass and stare into space, ignoring my hesitant heels in his flanks. While everyone else enjoyed the nature trail, I sat on Cosmo's back, my face getting scratched up by low-hanging branches.

The rest of the week we hung out at the condo. Aunt Carol bought a giant ham, which I nibbled on when no one was looking. I didn't want my aunt and cousins to think I was starving, for chrissakes. At night we gave each other beer shampoos with Coors, ate Ritz crackers and Port Salut cheese spread, played round after raucous round of Yahtzee.

"Mama needs a new pair of shoes!" Aunt Carol cackled, spitting on the dice and tossing them out with an enviable, effortless *joie de vivre*.

The Mountains made me appreciate the simple things. It didn't cost anything to go out and sit in the creek behind the condo with a pensive look on my face, waiting for someone to notice me sitting there and ask what brought a young girl out into the heart of nature to contemplate.

"I'm going out back to the creek," I announced. "Don't anyone bother to come look for me!"

The water in the creek was freezing cold and my legs went numb almost immediately. But I saw something glinting in the rocks on the creek bed. My face flooded with hot hopeful blood.

Gold! It was gold, pieces of gold in almost every rock! It would be hard work digging out the tiny flakes and nuggets. I puffed out my chest and reminded myself that hard work was the only kind of work worth doing. I would do this hard, backbreaking work with a hearty, nature-loving look on my face. Once I had extracted enough gold to buy myself a proud, galloping stallion named King Morningstar who would make that stupid Cosmo look like a piece of shit, I could use the rest to buy a mansion and pay for my parents' divorce.

Using my shirttail as a container, I gathered up as many of the stones as I could and ran back up to the condo.

"Look! GOLD!" I said, emptying the glittering stones onto the kitchen table, where my mother and Aunt Carol sat not talking and eating ham.

"Ha, that's fool's gold," Aunt Carol said, stubbing out her cigarette. "Look, I don't know about the rest of you, but I've had just about enough of sitting around on my tushie. Let's blow this joint! We're going out for dinner, my treat."

"Really, Carol, there's no need—there's so much ham," Mom began.

"Sister," Aunt Carol said, holding up her hand, "don't even bother to start with me."

We ended up at the restaurant in the Holiday Inn down the road, the result of some sort of compromise. Wendy and Lauren happily ordered lavish entreés, side dishes, and desserts galore, while my sisters and I insisted that really, we weren't all that hungry and would just split the hamburger four ways. Furthermore, we were truly delighted at drinking water—yes, just plain water—with our meal.

"Aw, cut the martyr act, Sis," Aunt Carol said, holding a drink and cigarette in one hand while she ate with the other. "Let 'em order Shirley Temples—I'll pay."

"Yeah!" I said, wanting to believe that when someone offered you something, they really meant it. "C'mon Mom, let her pay!"

I could almost feel the grenadine dribbling over my lips, taste the maraschino cherry on its tiny plastic sword.

"Oh Carol, you know you don't want to go doing that," Mom said, crushing my dreams. "Besides, carbonated drinks give you urinary tract infections."

My aunt ordered another drink.

When we went to leave the restaurant, Mom asked her for the car keys.

"What? I can drive!" Aunt Carol said.

"Carol, I don't think you should get behind the wheel," Mom said, reaching out her hand.

Behind the Wheel. I liked the sound of that: desperate, but still in control.

"You always think you know better than me," Aunt Carol shot back, stumbling over a curb in the parking lot.

"You are in no condition to drive these children," Mom said.

"Stop it! Stop fighting!" Wendy cried. "I want to ride with my mother."

My aunt and cousins got into their Country Squire wagon and peeled out.

"Shall I call for a taxi?" I asked, warming up my British accent, figuring a place like Vail employed a fleet of handsome cabs replete with uniformed liverymen.

Mom stared out after the swerving station wagon.

"We'll walk. It's not so far."

Walking along a gravel shoulder at night gives you plenty of time to watch the scenery go by.

When we got back to the condo, everything was dark.

"Hello?" Mom called, turning on the kitchen light. "Carol? Girls?"

"Well, there. They. Are," a voice hissed from above.

Hanging over the railing of the sleeping loft were my aunt and her daughters. Aunt Carol held the ham in her arms.

"Think they're too good for us, don't they?"

I don't think that! I wanted to scream. *This is all a big misunderstanding! My mother is stupid!*

I wanted to race up the ladder to the cozy loft, to snuggle down with them, to sing songs with them, to eat some ham with them. Then we'd all put on our dirndl skirts, embroidered vests, and huarache sandals, and I would lead them over the Mountains away from the Nazis.

The next morning, wordlessly, Aunt Carol drove us down to the bus depot.

Mom was right about one thing. The Mountains had dried up my mucus. The membranes inside my nose were cracked and bleeding. The bloody Kleenex plugs hanging out of my nostrils weren't going to be much of a man magnet for her or me.

I gave my fellow passengers the once-over. It appeared once again I had the whitest teeth on board. That settled, I was free to spend the return trip agonizing over what Dad would do to us when we got home. He'd probably already killed Scooter, letting him die of starvation in his little neon tiki hut.

We walked home from the bus depot through a thick curtain of rain. Drenched, we let ourselves into the house. It smelled different. Cleaner.

I rushed to the fishbowl on the kitchen counter.

"Scooter's still alive!" I said. "Dad didn't kill—"

Dad came through the doorway right next to me, making me jump.

"Well, look what the cat dragged in," he smiled, squeezing my shoulder. "Missed you guys."

He took his time, gently touching me and each of my sisters. We nuzzled him back.

Mom stood apart, on the other side of the kitchen table. Her nostrils flared, untrusting. Water dripped from her hair.

Dad made his way over to the table, resting his palms on the edge of its scratched faux-bois surface. The hair stood up on my arms as he leaned toward my mother.

"Next time, honey, before you go out rattin' the roads?" he said. "Do me a favor, and keep your mitts out of my pocketbook."

Choke

Nicole Sharp blew into St. Tammany Parish from someplace exotic like East Texas, and all the boys in seventh grade fell in love with her. She was tan and freckled and had pink shiny spots on her nose and cheeks, as though she were forever peeling from a recent sunny vacation.

Right away, I jumped into the New Girl Gap, that brief period of time before she realized her potential for popularity was much greater than mine.

"Hey Nicole, can you spend the night at my house Friday?" I asked. "Saturday my dad might make us rake the yard, but we can burn the pine needles afterward."

Might as well be honest from the get-go. I found the promise of fire usually compensated for a few hours of hard labor.

"Umm, I think I'm going to the movies Friday, but thanks anyway," Nicole smiled.

Guess I missed the gap. I walked over to Robin Powers, who was changing the rubber bands on her braces and pressing orthodontic wax onto her molars.

"You wanna shleep over Friday?" Robin slurred, swallowing the surplus saliva her headgear generated.

"I guess," I said, with minimal enthusiasm.

Friday afternoon I went home from school with Robin. We played Rummy-O for two hours.

"What movie are you going to see?" Mrs. Powers asked.

"Uhhh, I think we're gonna go see *Every Which Way But Loose*," I said.

I was startled by her sparkling, interested tone and by the question itself. Going to the movies never meant going to see a specific movie; seeing the movie was secondary to being seen.

Robin's mother was a former stewardess: lithe, angular, and flirtatious.

"Oh come on, that old Clint Eastwood movie?" she teased, sitting on my knee. "Aren't I prettier than he is? Why don't you stay in tonight with me? Pretty please?"

Robin wanted to, but I stood my ground. We would eat dinner with her family and nothing more. That was tough enough. Eating dinner at Robin's house was a painstaking affair. I had to pace myself, or I'd be looking over the edge of the plate I was licking clean to see Mr. Powers lifting his first forkful of spaghetti.

The Powers family were a wholesome lot. Robin and her brother Ricky played board games with their parents. They cuffed each other gently on the shoulder. They snuggled. They had HBO and Cinemax AND the Movie Channel, but Robin and Ricky weren't allowed to see R-rated movies. Whenever I insinuated that we sneak and watch, Robin shook her head.

"My parentsh trusht me," she'd say.

That night Robin's mom dropped us off at Cinema II in Mandeville. I strode around the parking lot, feeling pretty fine, mingling with the other kids before the movie started.

There was Nicole! Wait a minute. She appeared to be out on a date with Reggie Filbert.

"Hey, ugly!" Reggie said, clapping me on the back as though I deserved to be congratulated.

What was he talking about? I was wearing my best t-shirt, the one with the glittery Yosemite Sam iron-on custom-made at Print Me Please in Lakeside Mall. Okay, so the brand-new boot-cut Wranglers from Tally's Feed-N-Seed were a little too stiff, and the light green hooded sweatshirt across my shoulders was missing the drawstring. My scuffed mock-Tretorn sneakers were a little more mock, less Tretorn. I noticed that Yosemite Sam had a purple stain on his mustache like he'd been drinking Kool-Aid. He looked up at me from the flat plane of my chest, glittering. *Yew look like shit, varmint!*

"Reg-gie," Nicole said all sexy, smiling at me. God, she was benevolent beyond her years.

It occurred to me that standing next to Nicole wasn't doing me any favors. Her sprouting breastlets strained against the red plaid of her blouse, a magical fabric shot through with metallic gold thread. She and I were like a before and after picture for puberty.

I quickly found Robin. She was even skinnier than I was, and her chest wasn't just flat, it was concave. After we studied them in geography, the boys started calling her Crater Lakes. Robin and I hung out a while longer in the parking lot, watching some public school kids get into a head-butting contest.

By the time we settled in to watch *Every Which Way But Loose*, I was still trying to shake Reggie's *Hey ugly* out of my ears like pool water. I concentrated on concentrating on the movie.

Clint Eastwood played Philo Beddoe, a man who travels around the country in a jerry-rigged pickup truck camper, bare-knuckle fighting for cash. He lives with his mother, played by Ruth Gordon, his greasy, dim-witted cousin, and an orangutan that punches people and shits in the laundry basket. Philo's life was ugly and dirty. Even his love interest, a bar singer played by Sondra Locke, looked a little hard and drawn around the mouth. By the end of the movie, I felt beautiful again.

✲ ✲ ✲

Nicole was the star attraction at recess, performing endless windmill flips on the monkey bars. Sometimes she'd pause with the bar between her legs, telling a joke and laughing in that sexy hoarse way she had. She mesmerized us all, especially Mrs. Guidry, the teacher who got stuck with recess duty, P.E., and anything remotely physical, because she was one of the few faculty members who didn't smoke.

"Oh Nicole, I could barely pay attention to the judge last week when I was on jewelry duty," Mrs. Guidry said. "I just missed you so ba-yad!"

I wanted to know more about this so-called jewelry duty, but Nicole stepped forward and touched the Band-Aid on Mrs. Guidry's upper arm.

"That mercurochrome?" Nicole asked about the red medicine peeping out.

"Oh dang, I wish! The doctor's making me use Merthiolate and it burns like a—well, it burns!" Mrs. Guidry said. "Really Nicole, you should see the needle they use to break up a calcium deposit. It's huge!"

Nicole shook her head from side to side like a beautiful, caring horse.

One crisp March afternoon during afternoon recess, Nicole climbed up to the top rung of the bars and started spinning around fast, just a blur of glossy black hair and tan legs. I was sitting on a rusty old fifty-gallon drum with my chin on my fist, watching her dreamily. Nobody could windmill flip like Nicole. I took gymnastics Tuesdays after school, but the trampoline was more my milieu. I'd have mastered it by now if it weren't for Bruce.

Bruce was the soigné trampoline instructor at my gymnastics school. He wore black horn-rimmed glasses secured to his head with an elastic strap.

"Allow me to demonstrate Swivel Hips!" he'd announce, leaping onto the trampoline in his sweatpants with no underwear on beneath. "And SWIVEL, SWIVEL, SWIVEL, SWIVEL, SWIVEL, SWIVEL . . . "

Tramp Hog. I was lucky if I got a turn at all.

Watching Nicole turning and turning and turning on the bar, my eyes began to swim. I was almost in a full-blown trance when I heard something go CRACK!

"Owwwwww!" she yelled, stopping upside down in mid-flip, hanging from the bar by one knee.

Nicole's arm dangled down from a spot somewhere between her elbow and her wrist. It looked like the backward joint in a bird's leg.

"Nicole! Oh, Sweet Jesus, NO!" Mrs. Guidry scooped her up and ran toward the school building, a crowd of kids following behind like a funeral procession.

A hush descended on the campus.

Hmmmmmm. When I pretended to have a heat stroke to get out of a boring kickball game, I was collapsed in the outfield of the baseball diamond a good ten minutes before someone noticed.

"Sarah Thyre, you get up right now!" Mrs. Guidry called from the dugout, a dozen yards away. "That's no place for a nap!"

I lifted my head weakly and answered, "I'm not sleeping . . . I, I, I think I had a heat stroke . . . y'know, like you taught us about in health class day before yesterday?"

When Nicole broke her arm, it was like the Kennedy assassination. Everyone had a story about where they were when they found out. I was only two feet away at the time, so my eyewitness account trumped them all.

Nicole came back to school in a few days with a cool cast on her arm. Reggie Filbert drew a big magenta heart on it. It inspired enormous envy in all the girls, especially Tanya Tanner.

"I'll give y'all this pack of Starburst if y'all'll jump up and down on my leg 'til it breaks," she told some sixth-grade boys at recess the next day.

Tanya stretched out her leg, placing her ankle in a swing, and put her hands on her hips expectantly. Travis Kehoe and Marcel Toups began jumping on her freckled, fleshy thigh.

"Ouch! No—harder!" Tanya barked. "Owww! No! No, don't *stop*—'No' doesn't mean *no*, stupids!"

Marcel and Travis did their best, but the soft, supple cartilage of childhood and a nosey teacher who saw what was going on and decided to put a stop to it, kept Tanya from hearing the delicious *crack* of her femur splitting in two. When she ran out of Starburst and tired of slamming her arms in the heavy front doors of the school, Tanya gave up. I hated to see her so down on life.

I found a second grader with pink eye. It was a lovely case. Her right eye was crusted shut and the lid looked like a bright pink pair of wax lips.

"Gimme some of your eye juice," I said, dipping my fingers into her eye, scraping up a little of the crust with my nails.

"Tanya! Look what I've got!" I said, luxuriantly rubbing the juice first into her eyes, then into my own.

I was hoping we could hang out, maybe hold hands, and wait for our pink eye to blossom.

"I'm just gonna let this cook for a while," Tanya said, lying back on the bleachers and waving me away without a word of thanks.

She was out for two weeks and even needed to go to the hospital twice for emergency drainage. The whites of my own eyes stayed the color of brand-new tube socks.

✵ ✵ ✵

One stormy day all the seventh graders were herded into the cafeteria for P.E. The tables were stacked against the walls to make room for gym mats on the floor. Today we would be doing the sit-ups portion of the Presidential Physical Fitness Program.

I was a whiz at sit-ups. I held the class record of eighty-seven in a minute. They were completely non-regulation sit-ups: someone sat on my feet and held my knees and I rocked up and down using my tailbone as a fulcrum, propelled by momentum and nary a stomach muscle.

According to its mission statement, the Presidential Physical Fitness Program was meant to inspire a sense of healthy camaraderie: "Hey kids, let's all get fit—TOGETHER!" Instead, it brought out the cutthroat competitiveness in the athletic kids and left the fatties and spazzers crying in the corner.

Sure, I felt momentarily sorry for Nanette "Fatzit" Bouquet and Clark "Worm Dick" Lacoste, but pity was no match for winning. It was the one time a year I excelled at anything physical. I had to make the most of it.

"Sarah, can I hold your legs?" asked Denise Patrick, a rather unpopular, snaggletoothed girl whose own legs were dotted with scabs.

I glanced around for a partner of higher quality but then figured it was probably the job Denise was born to do, and the least I could do was let her do it.

As she clasped her arms and legs around my calves, I breathed a sigh of relief that I had worn tights that day to protect me from her leg disease, whatever it was. We waited for Mrs. Guidry to blow the start whistle.

Tweeeeeeeeeeeeee!

I breezed through my set, rocking furiously to and fro like that kid in the made-for-TV autism movie, *Son-Rise*. When the stop whistle

sounded, I hopped up in one fluid motion, prepared to garner awe and accolades for my new record of ninety-four.

"Wow!" Denise said, her front tooth completely encased in a boogery-looking shell.

Where was everybody? I squinted nonchalantly at the horizon and suppressed the urge to spit.

There everybody was, in a huddle down at the other end of the cafeteria. Denise ran over and I followed, flexing my arms and pausing for a deep knee bend or two.

It was Nicole. Her face looked almost purple, but still cute, and she was breathing all huffy. Our classmates were clustered around her, their hands reaching out to pat her or pet her or perhaps touch the hem of her garment.

"For God's sake—I mean for Pete's sake," Mrs. Guidry screeched, "back up now and give her some room! Nicole, Nicole, Nicole! Can you speak to me, honey?"

"What's happening?" I asked, as I studied my nails.

"Oh Sarah," Denise cried. "Nicole's having an azmertack!"

"Asthma attack, uglies," said Reggie Filbert, shoving us aside. "Looks like the two uglies finally got together."

"Oh, we're not together," I said.

"You're both on that fake cheerleading squad," Reggie said. "The one you gotta pay to be on."

Before I could explain how I had only joined the squad to lend it a little credibility and sex appeal, Reggie was gone, rushing to Nicole's side.

Denise blinked. Her mouth turned down at the corners and trembled in the middle.

"What are you crying about?" I said to her. "Nicole doesn't even like you."

I elbowed past her, trying to get as close to Nicole as I could.

"Really, I'm okay," Nicole coughed.

"Oh Nicole, leave it to you to say that! You are an absolute doll, I swear to Christ—I mean, I swear," Mrs. Guidry said. "Here, put on my jacket, baby."

"You're gonna pull through this, Nicky!" I called.

From then on, Nicole was the teacher's assistant during P.E. She was excused from doing any of the exercises. She even got a special patch from the Presidential Physical Fitness Committee, orange with white stripes.

I wasn't sure what an asthma attack was, but it had a dramatic ring to it. More important, it looked easy to fake.

The next day the rain had stopped and we were out on the monkey bars doing the flex arm hang. Normally I could hang all day, thanks to my natural scrawniness and Coach Brock's potbelly. He stood very close to us and allowed us to rest our knees on his stomach. Well, it was more like resting our crotches on his stomach, but there was nothing untoward about it.

The flex arm hang was booooooorrrrrrring to do and even more so to watch. I decided to spice things up a little. I sprinted around the monkey bars and collapsed in a heap, gasping and crawling through the mud. I gasped and crawled around my classmates twice before someone noticed.

"What's your problem, girl? You got the grippe?" Coach Brock asked me. "The cold ground ain't no place to be."

"I think—*wheeze*—I'm having—*cough*—an asthma attack," I panted, gouging my fingers in the mud like Charlton Heston on the beach at the end of *Planet of the Apes.*

I glanced at Nicole, who was holding the stopwatch and the clipboard with everyone's hang times. She looked at me skeptically and tapped a pen against her teeth.

"Well, go get yourself a drink of water, then," Coach said, turning back to the girl resting her mons veneris on his belly.

Well, a drink of water was something out of the ordinary. I'd take what I could get.

I decided to try out this asthma thing on my mother. Already, practically every morning at breakfast, she'd purse her lips and cock her head and say, "You look pale. Are you sure you feel up to school?"

She let me stay home almost as often as I liked. How nice it would be for both of us to finally have a legitimate medical excuse.

Did asthma come with fever? I wasn't sure, but I pulled the old thermometer-on-the-light-bulb trick just to seal the deal.

"Well, your forehead feels pretty cool," Mom said, but she couldn't resist. "You better stay home."

After a few weeks of delicate sighs and shallow coughs and five or six sick days, something changed. I began waking up in the middle of the night with deep, prolonged coughing fits that pulled so hard at my abdominal muscles I peed my pants. Mom would help me out to the sofa and prop me up with pillows. There I slept, fitfully, until the sun came up, my chest puffed up tight and the sound of my breathing a whiny, whistling rasp.

Somewhere along the line, my asthma had turned real.

We went to see Dr. Camp. He prescribed a grey inhaler, "two puffs, three times a day." It shot out a cloud of metal-flavored air that I found positively addictive. At school, I was thrilled to discover an inhaler had an exotic cachet. I welcomed any questions and honored all requests for a demonstration.

Pretty soon I was up to four puffs, five times a day. But I wasn't getting any better.

I felt taken advantage of. This wasn't supposed to be part of the deal. I had assumed asthma was high drama, low pain. I just wanted an easily controllable excuse to get out of P.E. I wanted an accessory, not the whole outfit. What was happening?

Dr. Camp recommended a pricey specialist. The cost of my many appointments enraged my father.

"Another god DAMN sixty-dollar office visit?"

Mom would lead me out to the car slowly as I wheezed.

"You are my suffering little angel, Sarah."

The specialist, a tiny Indian lady named Dr. Singh, was no romantic. She withheld any sliver of bedside manner as though it cost extra and we couldn't possibly afford it. I got the feeling from our very first visit she knew we were the collection agency type. Every time I went to see her I'd cry, which only made breathing more difficult.

"Seedah, I von't hoordt you," Dr. Singh stared coldly at me from across the room.

She gouged eighty scratches on my back with a razor blade and applied a drop of allergenic fluid to each.

"Djoo are allairgic to every grass, every pollen, and every animal except sheep," Dr. Singh reported after looking at the welts on my back. "Djoo are also highly allairgic to mold. Djoo need to rid your house of mold."

Mom and I looked at each other, thinking the same thing: scrub as hard as we might, that slimy black coating on the shower wall wasn't going anywhere. My hope for a quick fix was evaporating.

"We need to do further testing. I will be giving you twenty shots in the arm, Seedah," Dr. Singh said with a sadistic smile.

"Noooooooooo!" I cried.

"Don't wuddy, they are only skin-pops, like heroin addicts do," she said, providing me with the proper lingo in case I were inclined to become an IV drug user.

The shots made me throw up all over her office, but it was worth it. She determined that I was exceptionally allergic to horses. Now I had a convenient excuse to opt out of the cost-prohibitive riding hobby that was becoming all the rage among the rich girls in my class.

I got jittery on the various pills and inhalers Dr. Singh prescribed. Sleep left me almost entirely. When I had an attack, I didn't turn that lovely shade of Nicole purple, just a ruddy, poor man's red. I developed dark circles under my eyes. I lost even more weight. I looked like a ghoul.

If I managed to doze off propped up on the sofa, I woke up with Mom sprinkling holy water on me, under a hillock of prayer pamphlets, wearing so many rosaries, saints' medals, and scapulars I looked like Mr. T. A scapular is a cloth medallion imprinted with a picture of the Virgin Mary and a brief prayer for salvation. Catholics believe if you die wearing one, you are immediately sucked up into Heaven, no questions asked. Mom kept one in her bra at all times. Slowly it sank in that she didn't expect me to make it through the night.

"Seedah, djoor asthma is chronic and acute," Dr. Singh told me. "Djoo need allairgy shots."

More shots! Just the thought of them made me throw up.

"Braaaaaghhhh," I gagged, trying to make myself throw up.

"Shots von't hoordt you, Seedah," she stated flatly, pounding each syllable out on her desk. She showed my mother how to practice giving injections on an orange.

"Ohhhh, I don't think I can do this!" Mom said, turning her face away from the orange while blindly stabbing at it with the syringe. "What if I get an air bubble in there? Won't it travel up Sarah's bloodstream and lodge in her brain?"

"All dight, all dight," sighed Dr. Singh, staring at Mom and me with disgust. "Come in tomorrow afternoon and the noorse will administer the shot. But you have to keep the seedum in your own refrigerator. We are not running an apartament house here."

Dr. Singh handed us the Styrofoam box containing the vials of serum: death by lethal injection.

There was nothing left to do but run away from home. The next day at school, when the dismissal bell rang at 3 o'clock, instead of boarding my usual bus, I crept across the parking lot and snuck into an abandoned one. Lying low on the bus floor, I killed a half an hour doing my homework. Then I had to do my lines.

Like nuns, corporal punishment had gone out of style. The laypeople preferred giving lines. Mrs. Voorhies, the school librarian, had overheard me telling Reggie Filbert to shut up. How was I to know a librarian despised the phrase "shut up?" She ordered me to write "I will not say shut up" five hundred times.

Mrs. Faulk had given me lines that day, too. On the playground at recess, she intercepted a petition I was circulating against Bitsy Marshall, whose crass materialism had clearly gotten out of control.

> *We the undersigned find you to be a SNOB. We don't care about your alligator shirts or your new Calvin Klein jeans or your seven pairs of Jacques Cohen espadrilles in a rainbow of colors or the new 16-millimeter lapis lazuli on your Add-a-Bead necklace.*

I was getting signatures when Mrs. Faulk came up and snatched it out of my hands.

"This is disssgusting," she sibilated. "Though I'm not surprised. Typical girl behavior. Sssarah, I'm giving you lines for this. You will write, 'I will not start petitions againsst people' five hundred times and turn it in to me tomorrow."

I thought that was a little extreme. I'd been led to believe that petitions were a valid tool for change and progress. I considered petitioning Principal Lusco for the right to petition but somehow knew the exquisite subtleties of my argument would go unappreciated.

Crouched on the bus, I wrote my thousand lines. I got up and stretched just in time to see Mom drive through the parking lot in our Gremlin. I fell to the floor and crawled back toward the emergency exit. I wasn't sure what time it was, but Dr. Singh's office was probably still open for business.

I'd been on this bus a couple of hours at least. Long enough to have to go to the bathroom. I peeked at the school. It looked closed for the night. Still too light outside to risk going outside.

I went on a piece of loose-leaf paper in the back of the bus, wiping neatly with the cleanest-looking thing within reach, an empty Wacky Wafers wrapper.

Finally getting dark. Time to move on. I pulled the lever that opened the bus door. Always wanted to do that. I opened and closed the door ten or thirty times and then hopped out and ambled about, kicking the crushed oyster shells of the parking lot. It felt good to be moving around and making some noise after being cooped up for so long.

I moseyed over to the church, figuring my mom might be in there, praying for me. I stuck my head in the back door and viewed the small weeknight congregation. It was Wednesday: Stations of the Cross. Father Adrian was leading a group of people around the church, stopping in front of each picture in the series of fourteen that detailed Christ's ascent up Calvary to his crucifixion.

Father Adrian was just starting the prayer for the Sixth Station. I loved the Sixth Station. That's the one where a gal named Veronica presses her hankie on Jesus' face to wipe off the "vulgar blood and mocking spit."

Veronica had glossy black hair like Nicole's. For her trouble, she gets sainthood, not to mention an indelible portrait of the Savior's face—in his prime—for a souvenir.

I closed my eyes and listened to the prayers until Father Adrian moved on to the Seventh Station: Jesus Falls a Second Time. A little repetitive, not really my cup of tea. I opened my eyes and scanned the faces of the churchgoers.

Something lashed out and wrapped itself around my neck like a tentacle. Another tentacle latched onto my wrist. Next thing I knew, Sister Titus's mean monkey face was in mine.

"Where the hell have you been?" she stage-whispered.

I pictured fumes wafting off the telltale turd on the school bus.

"B-b-b-behind the bleachers," I stuttered.

"You little liar!"

I decided to change the subject.

"Hey, Sister Titus," I said ultra-casually. "Seen my mom around here?"

"If I was your mama, I wouldn't be praying, I'd be cursing you right now!" she hissed, tightening her vise grip on my neck.

Sister Titus hustled me over to the convent foyer and called my house. Mom arrived minutes later, crying and furious.

"I thought you were kidnapped, raped, mutilated, dead, and rotting in a ditch somewhere!" she sobbed in the car.

"I certainly hope you're going to report Sister Titus to Principal Lusco," I said. "Look at my wrist—she practically gave me an Indian burn!"

The serum for the allergy shots sat in our refrigerator long past its expiration date. Mom never again mentioned putting it into my body. She had decided that Dr. Singh's brand of Western medicine wasn't the right mode of treatment for my asthma, and I wasn't going to insist upon a strict schedule of injections.

"We're going up to St. Anne's tonight. There's a band of Charismatic Christians in town. One night only!" Mom said one evening. "Debbie, you come too. You had all those kidney infections last year."

Mom had lost faith in Western medicine for Debbie, too. The last time she'd taken Debbie to the urologist, Mom peed in the cup herself for Debbie's urinalysis. The doctor himself called, alarmed at the blood and hormones in the urine.

"Well, somebody had to do it," Mom said into the telephone. "Poor child was too tired to pee."

Debbie hadn't had any kidney trouble lately, but a little proactive prayer couldn't hurt.

When we got to St. Peter's, we stood in the back during the ceremony, embarrassed because it wasn't our regular church and we'd just come for a miracle. I couldn't see very well from the rear of the church. Up on the altar a group of heads was singing and moving around in a choreographed way. There was much bell ringing.

When it was over, Mom approached their leader, an old guy who looked like someone had taken Vincent Price and whittled him down to half his normal size. She whispered something in his ear, pointing over at me. He nodded and cracked his knuckles.

Little Vincent walked toward me, arms outstretched and palms splayed. Before I could run, he clamped his hands down firmly on my shoulders and threw back his head.

"Sha-la-la-la-la-la, sha-la-la-la-la-la," he chanted in an eerie monotone.

Mom smiled beatifically, swaying along to the rhythm. I was mortified but knew it would be rude to squirm. I snuck a glance over at Debbie. Some other guy was doing exactly the same thing to her. Debbie looked straight ahead, her already huge eyes positively bulbous.

All this was happening in the entrance hall to the church, where the disbanded Charismatics were shuffling around, drinking coffee and eating those star-shaped cookies daubed with a little smear of red jelly. Every once in a while, a few of the Charismatics would pause and appreciate the spectacle of healing, smile knowingly at each other, and then return to their chatting and mingling.

Finally, Little Vincent released me from his grip, pushed me forward about five feet into the cookie-eating crowd, and collapsed into a chair. A woman ran up to wipe his brow with a handkerchief.

"You're healed, little girl," she chirped, popping a cookie into Vincent's mouth. "And lucky for you, you've been healed by the best!"

Having got what we came for, Debbie and I skulked out of the church, dragging our exuberant mother along.

"Well, that was fun!" Mom said. "Maybe we should start coming up to St. Peter's for Mass on Sundays. These people are friendly."

"I can't believe you tricked us into this!" I screamed. "I don't want to go to this church. Those people are weird!"

"That man was speaking in tongues," Mom explained. "He could put a bug in God's ear about you."

"I don't want him to!" I said, revolted by the image of a giant beetle scuttling into God's hairy old ear.

"Oh, you've just been slain in the Spirit," Mom said calmly, watching the road. "You'll be thankful when your asthma's gone."

Much to my mother's puzzlement, my asthma didn't go away. In fact, it got even worse, turning into a sort of chronic bronchitis that settled deep in my chest and racked my body with convulsive coughing. It was enough to drive Mom back into the arms of science. Halfway.

"I heard about a doctor who's good and cheap," Mom said, drinking her seventh cup of coffee after a long, sleepless, wheeze-filled night. "With his help, I think we can lick this thing once and for all."

We drove over to his office. The nurse led us into an examination room. My suspicions were immediately aroused by the bleeding Sacred Heart of Jesus poster taped to the wall. Before I could kick Mom in the shin, the doctor came in.

"Hello . . . Sarah, is it? I'm Dr. Gremillion," he said, his manner mild as a priest's. "In just a moment, I'm going to perform the laying on of hands."

"I'll just be out in the waiting room," Mom said, ducking through the door.

Mom wasn't taking any chances. A month ago she'd passed out watching Dr. Camp burn a bleeding wart off Becky's left ring finger. Seeing her flee made me certain something gruesome was about to happen to me.

Dr. Gremillion's arms shot up in the air and froze into a V. His head fell forward, spent. He looked like Judy Garland at the end of a big number.

"Okay!" I said, expecting quite a show.

Slowly, his arms drifted down. His hands landed on my neck so gently it felt like a necklace: a beautiful, shining, hammered-gold breastplate resting upon my collarbone. My neck lengthened. I felt taller, and proud, with a high forehead like Nefertiti's.

"Hmmmmm," he said. "I see."

"What?" I said, my voice sounding harsher than I intended.

"Ssshhhhhh," Dr. Gremillion said, palpating.

I'm not sure, but I think I went to sleep for a few minutes.

"Sarah, Sarah," Dr. Gremillion said, shaking me. "You have a lot of anger toward someone, Sarah. If you forgive that person, your asthma will disappear, Sarah."

Pressing a little harder on my chest, Dr. Gremillion wheeled gently back and forth in his swivel chair.

"O Lord!" he shouted. "Take it from her, Lord, this anger! Release her from it, Lord, we ask with humility!"

His hands flitted up from my body, landed, and then flitted away again in a fanning motion.

"Go 'way, asthma!" he whispered. "Go 'way!"

His hands landed back on my chest, heavy.

"Lord, give her the strength to make this sacrifice," he said, his chair creaking to a reverent stop.

There was a long pause.

"Well," I said. "That was great."

"Was it?" Dr. Gremillion asked. "Was it really?"

"Sure!" I said, patting his hands, which were still on my chest.

Dr. Gremillion went over to his desk and began writing something.

"Don't forget the sacrifice," he said, handing me a slip of paper.

It was a prescription for antibiotics.

On cue, Mom stepped back into the room, smiling her saintly smile. At first I wanted to kill her. Then I thought about what the doctor said about being angry with someone. Okay, I wouldn't be mad. I was tired of having asthma and wanted it to go away. If I had enough faith or serenity or whatever it took, maybe it would go away. I had to believe this would work, or it wouldn't work, right?

Mom and I got back into the car and started to drive home. I folded my hands in my lap and looked out the window, admiring my new serene countenance in the dangling side-view mirror. I was calm. I was cured.

"By the way," Mom said, reaching behind her seat and pulling out a jar, "you know that man who sells shrimp out on Highway 190? The guy with the sign that says 'BIG SWIMPS,' and the snake tattoo that wraps around his neck and up onto his forehead? His name's Frank. He's raising bees, now, and selling the honey. It's all-natural. Frank says if you eat a spoonful a day you'll inoculate yourself with pollen and build up some immunity and your asthma will probably go away."

I held the jar up to the light. Twigs and leaves were suspended in the dark tar of the honey. I caught sight of my face in the mirror. The serenity was draining out of it.

"Just in case all that healing didn't take," Mom said, patting my leg. "Better to be safe than sorry."

Rubbing Alcohol and Vaseline

"What's a douche?" nine-year-old Hannah asked, lying on the floor in front of the TV.

Becky and I were draped across the sofa in the den, and my dad was a few feet away in his plaid recliner. Lethargic from Steak Diane, the sour magnum opus of Mom's gourmet oeuvre, we had converged for our weekly viewing of *Dallas*. A Massengill commercial just came on, the one where the woman looks wistfully out a window, yearning to be as fresh as a country lane after a spring shower.

"What's a douche?" Hannah asked again, turning around and raising her voice.

I should have seen this coming, after last week's stomach-churning "What's a call girl?" incident during *Three's Company*. I should have escaped to the bathroom, or out to the kitchen to whip up my homemade Doritos: Crunch up a bunch of taco shells in a Ziploc baggie. Add salt. Shake. Enjoy. Far more enjoyable than the other available snack options: raw oats or dry spaghetti noodles.

Dammit, why should I have to scurry at each commercial break? I might miss one of my favorites, like the one for the fiber powder that you can sprinkle over everything, even beef stew. Or the one for that chewy chocolate diet candy, Ayds.

"Ayds helped me lose forty pounds!" the lady in the commercial boasted, rolling a cube of it around in her mouth. "Mmmmmmmmm . . . Ayds."

I sank into the sofa as much as possible, which wasn't much. Our sofa was the kind with unforgiving foam cushions lying on an open wooden frame, more a padded bench than a sofa. Furthermore, I was the only one who called it a "sofa," which had a plush, classy ring to it. It hadn't caught on. Everyone else in the family said "couch," which sounded to me like a rash contracted by hicks:

> *Say, what's that on yer neck? A boil?*
> *Nah. Jest my couch flaring up again.*

Hannah stomped over to us. I felt Becky go stiff. Dad shifted in his recliner. I didn't dare look at him. If we were lucky, he'd dozed off. I squinched my eyes together and tried telepathically communicating.

SHUT UP SHUT UP SHUT UP.

My ESP told me Becky was doing the same thing.

"Didn't anybody hear what I said? I said, *What's a douche?!*" Hannah yelled, clearly lacking both intuition and common sense. "What's. A. DOUCHE."

Dad cleared his throat.

"Do you know what a VAGINA is?" he said, his voice like a hand clutching my heart.

"Yes," Hannah whispered, tears practically shooting out of her eyes.

"Well, ladies use 'em to clean out their VAGINAS. That's what a douche is," Dad said, standing up and clicking out of the room.

My father walked with a pronounced limp. When he was twenty-one, he went to buy a Triumph motorcycle. After signing the pink slip, he pulled out of the dealership parking lot and immediately skittered across Gregory Boulevard, breaking his ankle and shredding his leg.

"*It was hanging by a string,*" Mom would whisper, telling and retelling it like a ghost story. She had clipped the newspaper article about the accident from the *Kansas City Star* and put it in a tiny gold picture frame, which she gave Dad for his twenty-second birthday. He kept it in his sock drawer. I'd seen it there. Whenever Hannah and I felt that Dad had been mean to us, we got our revenge by spitting on his clean socks.

The doctors slapped the flesh back onto Dad's ankle in a shiny red-and-white pile, its surface forming a diagonal, the hypotenuse to his foot and leg. Every once in a while the knot broke out in sores that made it look like a mound of raw pork sausage. I helped prepare Betadine soaks and caught the ooze with gauze pads. I pretended to be one of the child laborers featured on the PBS show *Zoom:* living on a farm and tending to my horse Wildfire's infected hoof.

When Dad walked, his ankle joint emitted a rhythmic *clickle* that warned of his approach. It was like a metronome implant: the faster the tempo, the faster he was moving. My ears would prick up at the sound. I made split-second calculations, then hid or ceased whatever potentially incriminating thing I was doing, thereby avoiding being blindsided.

Becky and Hannah and I all held our breath until the clicking receded and the back door slammed.

"Stupid retard!" I sneered at Hannah, so upset I forgave myself the redundancy.

I jumped up from the sofa, shoved her down, and ran to my room, where I planned on crying until someone came looking for me.

"What's wrong?" Mom said, passing through a few minutes later to put a stack of clean clothes on my dresser.

"*Douche!*" was all I managed to sputter.

"Now, there's no need for that. The juice in your vagina is there for a reason. Your vagina cleans itself," Mom said, tucking my pillow under her chin and pulling a fresh case over it."Besides, douching causes cancer."

I still wasn't entirely clear about exactly how many holes were between my legs, but I got the feeling one of them was up to no good.

A few days later, while digging through the plastic bag Mom used as a purse, I came across a scrap of paper with a date and a curious-sounding name scribbled on the bottom of it.

"Soooooo, who's this *Pap Smear*?" I asked Mom, sure I was catching her having an affair. Pap would be an older man: bearded, courtly, possibly a jazz musician, definitely with a swimming pool.

"That's not a who," Mom said, "that's when they take a spatula and scrape out your vagina to see if you have cancer."

"Ucchhh!" I said, disappointed that I wouldn't be swimming at Pap's mansion and entertaining friends in his gazebo.

✵ ✵ ✵

I was out on the swings at recess, pumping ever higher and higher, lying waaaaay back and blinding myself with the sun's glare. I got the swing so high I was in danger of going up and over the bar. My vision was full of floating purple holes and blind spots. As I reached the apex of the swing's forward arc, I jumped out and flew through the air with my legs spread wide, crying out, "PAP SMEEEEEARRRRR!"

I landed in a heap at the feet of Michelle Banks, who sniffed and rolled her eyes.

"Jesus, Sarah, you practically like, maimed me," she said, rattling a little paper tube reproachfully.

"What's that?" I asked.

"Gawd! It's like, a *tampon*?" Michelle said, her eyebrows almost meeting her hairline. She had a very short forehead. "Jesus, don't you even know what a *period* is?"

Before I could answer, she traipsed off to the bathroom, brandishing her tampon far and wide.

Jesus, of course I knew what a period was. I'd read *Are You There God? It's Me, Margaret* and every other Judy Blume book I could get my hands on. I learned from *Then Again, Maybe I Won't* that boys masturbate, and from *Forever* that if you put aftershave on a guy's balls, he'll complain about it. But all those things happened to people in the books. They were all Jewish and lived in New Jersey. They went to temple instead of church, and when someone died, they sat around eating lox and bagels and half-sour pickles.

Becky had gotten her period, but she was old. She was in ninth grade. We hadn't bonded since the previous summer, when we both read *Helter Skelter* and were too terrified to sleep alone. Once the fear of being stabbed to death by total strangers abated, Becky withdrew into her own secret world, locking herself in her room to eat bag after bag of sunflower seeds and lip-synch Barbra Streisand's "Eyes of Laura Mars" into a hairbrush in front of the mirror. Occasionally, she would forget one of her used maxi-pads, bundled in a sheath of tissue, on the back of the toilet. Mom hated that so much she'd transfer the pad to the pillow on Becky's bed, like a vengeful hotel chocolate. It was one of the many things they argued about.

Thanks to Michelle Banks's little serving of snot, my eyes had been opened. Girls my age were getting their periods, too.

"Guess whose friend dropped by for a visit?" winked Malison Drucker.

"These darn cramps!" Bitsy Marshall moaned in reading class, clutching her lower abdomen.

"Does anyone have a tampon? I need Super Plus," Ellen LaCour bragged before P.E.

These girls had one thing in common. As they waltzed around the playground with their arms around each other, you could hear them proclaiming, "My mom's my best friend!" These girls talked to their moms about makeup and clothes and how to get finger-fucked. Mom would hardly condone third base when she already hated anything to do with second.

"Richard Gere shouldn't have licked Debra Winger's breasts like that," was her tidy review of *An Officer and a Gentleman*, which she and Dad had gone to see.

I would settle for a sanitized version. I could hover over her while she sat at an antique mahogany vanity, putting on red lipstick and spitting on cake mascara.

"Thank you, pet," Mom would say, as I fastened the clasp of her pearls. "Whatever would I do without you, my little darling!"

I would spritz some Estée Lauder Youth Dew on her pulse points and watch her float out into the night, smelling like the secretions from a frightened woodland creature's hind-glands.

First, we needed an antique mahogany vanity table.

I remember Mom being sexy once, back in Kansas City, circa 1975. She was wearing square-heeled silver sandals and a long green patchwork hostess skirt with a black turtleneck sweater: how chic! My parents were going out to a New Year's Eve party. Mom was forming a smoky cheese-ball by hand: molding together shredded sharp cheddar and softened cream cheese and liquid smoke, pressing chopped nuts into the outside. Dad came up behind her, grabbed her messy hand, and licked slivered almonds off her fingers. Mom laughed and smeared some cheese on his face, then licked it off. I ran away embarrassed, a regretful Peeping Tom.

But that was just about it. There was no "everyday sexy" like the book *Total Woman* recommended. Vicky Laschke had given my mother a copy.

"This'll really help you toot your hubby's horn!" Vicky winked, swatting Mom on the fanny with it.

Total Woman told housewives to be "feminine, soft, and touchable" rather than "dumpy, stringy, or exhausted." It said to "remove all prickly hairs" and greet your husband at the front door wearing a subtle, sexy costume, perhaps made of Saran Wrap.

We didn't even have Saran Wrap in the house! Mom preferred wax paper because it was biodegradable. How would she fashion a subtle, sexy costume out of that? Rustle, rustle, rustle . . . "Welcome home, honey!"

Why couldn't my mom be a little more like Tanya Tanner's mom?

"My mom's my best friend," Tanya gloated one day when I was over at her house. "We both use Frost 'n' Tips in our hair, and last Christmas she put tampons and douche and rubbers in my stocking."

"Just for laughs, sweetie, just for laughs!" Mrs. Tanner said in a caramelized voice, unsnapping a red vinyl cigarette case with her matching lacquered nails. She placed a fresh, long, brown More between her creamy pink lips and lit it against the old butt.

Other than Tanya's mom, the closest I got to sexy was the 4-H "Night of 100 Stars" at the Southeastern Louisiana Mental Hospital for Children. Sister Giles had us do a talent show for what she called "those p-p-poor distoibed children."

The orange-sequined dance costume Mom found at Goodwill inspired my improvised rock-jazz routine, set to "Greased Lightnin'." I swung those fringed armlets a hundred times backward and then a hundred times forward. Ninety-eight, ninety-nine, one hundred. I ran out of moves right around the line, *You know this ain't no braggin' / She's a real pussy wagon.*

I froze, in character, looking out at the audience. I noticed that several of the boys were biting their fingers and wrists. *Over me.* It would have been nicer without the pervasive scent of disinfectant and urine, but I wasn't going to grouse about it. I ended on the floor with my legs in the air, Busby Berkeley style.

Maybe I'd gotten my period and didn't even know it! I unscrewed my mom's makeup mirror and sat on the toilet. Okay, there was Pee Mountain; let's hitch that out of the way.

Hey, wait a minute . . . there were TWO holes between Pee Mountain and that dark pucker where poo came out. Surprise, surprise. That bonus hole in the middle must be the vagina. How the hell did a penis go into that? Not to mention a baby coming out of it?! It already looked full. Of skin.

"Mom, new Playtex Tampon Juniors have an Easy Glide Comfort Applicator!" I said, generously giving her an opportunity to be my best friend.

"Tampons make you aware of yourself, just like tight jeans," she said. "Why are you talking about this? Did you start?"

"Starting" was Mom's shorthand for getting my first period. Ever since I turned nine, whenever I was particularly emotional, sad, or angry, she'd chuckle maddeningly and say, "Oh, you're just getting ready to start."

"I'm not sure," I said mysteriously. "Maybe."

"I didn't get my period 'til I was sixteen," Mom said, dashing my hopes. "If you start, let me know and I'll buy you some granny pads."

For some reason she called maxi-pads "granny pads." So now I had the mental image of myself as an old lady, bent under the weight of my dowager's hump, pressing a pad into my underwear and crying out, "Eh? Whazzat? *What's happening to me?!*"

So much for trying to be best friends with my mom. Fine, I would get a job and buy the supplies myself.

✵ ✵ ✵

Our next-door neighbors, the Moores, were a middle-aged couple with two teenage daughters. The older one, Sandy, had two kids out of wedlock with different fathers. My parents were invited to the wedding when she married her new man, Tank. An hour or two into the reception, things weren't looking too good.

"Shit girl, I ain't got to tell you when I'll be back," Tank said, taking off on his motorcycle.

"Poor Mr. and Mrs. Moore. You should have seen their faces when Tank rode off with his buddies," Mom said. "I swear, if you ever get pregnant from premarital sex, don't come crying to me."

Duh. If I ever got pregnant from premarital sex, I would hardly go crying to anyone. Most likely I would call my boyfriend to warn him that my father was coming over with a gun, as in that song "Run Joey Run." Only I wouldn't die like Julie, the girl in the song. My dad would trip on his gun and die, and my boyfriend and I would laugh in his dead face, then ride off in an orange MG Midget convertible.

After her divorce, Sandy liked to go out every Friday night. I took the regular gig babysitting for Sean, three, and six-month-old Buster. They were always sweating out fevers in thick acrylic footie pajamas. They barely slept because Sandy gave them Coca-Cola in their baby bottles. Bill, the Moores' albino Irish setter, watched me all night long with two strings of saliva streaming from the corners of his mouth, taking the occasional break to scratch or lick his balding, pink rear end. Once Sean threw up what looked like scrambled eggs and onions on the living room carpet and before I could clean it up, Bill swooped down and ate it. He came in handy that way. The job paid a dollar an hour.

"Get those kids to sleep, babe, and you can work on my candle bottles," Sandy would say, shaking a box of lighters and matches at me and gesturing toward the dozen Chianti bottles, their necks spattered with colored wax drippings, on the tufted vinyl bar.

After a few weekends, I had enough money to buy a trial size box of New Freedom mini-pads. I snuck it to school in my backpack and placed it in a highly conspicuous place, hanging halfway out of my desk.

Five minutes into the first class of the day, English, I raised my hand.

"Yes, Sarah?" Mrs. Wilbury said, her jowls lending her voice a creamy resonance.

Mrs. Wilbury was over six feet tall and weighed at least 350 pounds. Hence, we referred to her as Mrs. Wheelbarrow Full of Lard. The other seventh-grade teacher was named Miss Fagot but no one made fun of her. It just went to show that, back in 1980, being fat was way worse than sounding gay.

"Mrs. Wilbury?" I said. "I need to go to the bathroom."

"All right," she said.

"I mean, I really need to go to the bathroom!" I said, picking up my box of New Freedom mini-pads, pointing at it with my index finger, then executing a half-turn at the waist with the box in my palm like Holly, the spunky, redheaded spokesmodel on *The Price Is Right.*

"I said fine," Mrs. Wilbury sighed, not even bothering to look up from her lesson plan.

I spun around on one foot, slowly enough to make eye contact with a few of the girls I knew had gotten their periods. Should I jump up in the air like the girl on the New Freedom package? The symmetry would be nice. . . .

Nah, too obvious. I just smiled and shrugged like, "What else can ya do when there's blood pouring outta ya? Am I right, gals, or am I *right*?"

✵ ✵ ✵

Summer approached. Ginger Talley and I volunteered to stay after school to put price tags on bottles of wine donated for the Wheel of Fortune booth at the school fair. I had no idea what the wine's fair market value was. My parents drank beer, scotch, and vodka. If the label had an animal or a flower on it, I jacked up the price to five bucks. It was almost 6 o'clock when we were done and went outside to wait for our moms on the bench near the flagpole.

Ginger flared her slitty little nostrils and said, "You know what? My mom says your mom's let herself go."

I'd never heard that phrase before. But I knew what it meant. Ginger's mom was a sexy divorcée. The one time I went over to Ginger's, her mom let us drink Champale and sit on her new boyfriend Rocky's lap.

Had my mother let herself go?

Mom pulled up in our old green truck, the used one my dad had bought from his company's retired fleet.

"No need to get out, Mom," I said frantically, running and jumping into the back of the truck. "Let's go!"

Too late. She hopped out wearing a paint-spattered men's t-shirt, cutoff jeans, and no bra. She walked over to say hello to Mr. Robert, the substitute teacher who looked like a cross between Colonel Sanders and Charles Nelson Reilly. He rubbed his hands on the front of his light aqua leisure suit, mumbled something noncommittal, and gazed out over the school parking lot.

Oh my god, it's true. Mom's long wavy hair was held up mammy-style with a dirty bandanna. And she looked FAT. Even Mr. Robert couldn't bear to lay eyes on her!

Oblivious to how repellent she had become, Mom clambered back up into the driver's seat and poked her head out the little rear window.

"We gotta stop and get gas on the way home," she sang. "Hang on!"

I lay down in the bed of the truck and hoped to die. I would, however, settle for remaining unseen. At the gas station, I got out to peruse the candy selection and pretended not to be with my mother.

"We'll take the special, Shawna," Mom said, chatting up the pregnant cashier. A sign outside promised a free pound of ground meat with every fill-up.

"How you been, girl?" Shawna said.

"Fine, just fine. How're you holding up? Due in July, right?" Mom asked.

"Yeah," Shawna said, taking a drag on her cigarette. "When you due, girl? October, huh?"

"Yep," Mom said.

I was shocked. As Mom pumped gas into the truck, I got hysterical.

"HEY—are you pregnant? Are you pregnant *again*?!" I said, trying to sound like the woman on *Fantasy Island* who found out her husband was a vampire.

"Yes," Mom said, as though she hadn't been hiding it, she'd just been waiting for someone to ask.

I had a bad feeling about this. Why had she told a cashier at the filling station she was pregnant, but none of us in her immediate family? As far as I could tell, Dad hadn't noticed her bulging stomach. Not surprising since Mom spent most nights on the couch. Sofa.

"Don't tell anyone," Mom said. "Your father doesn't know yet."

Dad found out pretty soon, and he was not happy.

"Whose is it?" he asked. "Mike's?"

Mike was our garbage man. When I heard Dad say that, I got more excited about this whole pregnancy deal. I couldn't wait for my adorable biracial half-sister. She'd have a creamy Lady Marmalade complexion and a wild, tawny Afro that I could pick out for hours. I'd always wanted to keep a pick in my back pocket. It just screamed, "Casual!"

Right away, I called all the girls in my class to tell them don't worry: my mother was NOT fat, she was pregnant.

"Thank god, I was beginning to wonder if she'd let herself go," said Ellen Lacour. "Oops! Gotta run—I just got my period."

Off and on, whenever Mom could scrape together enough money to pay for a class, I still took gymnastics up at Covington Athletics. One day we spent the whole hour straddling a wide vault, practicing dismounts. I went to the bathroom when I got home and there was a spot of blood in my leotard.

"*Mom!*" I hollered like a Judy Blume character, bracing myself for the most magical moment of our lives. "I got my period!"

Mom came into the bathroom all distracted.

"Lay down on the bath mat," she said.

I got down on the damp rug that always smelled like wet dog.

"Spread your legs wide so I can get a good look," she said.

My heart was beating fast. I felt so grown-up all of a sudden.

"Well, I don't know what that is—but that isn't a period. Maybe you ripped yourself when you were doing the splits or something. Wash it with rubbing alcohol and put some Vaseline on it."

This one-two punch was our family's cure-all for everything, be it abrasion or deep puncture wound. When the handlebars rusted off my bike while I was riding downhill and I careened into a ditch, Becky had dressed my gravel-studded scrapes this way. Apparently, it would heal a ripped-open vagina, too.

I sat on the toilet and poured a little alcohol on a wad of tissue. The fumes alone set off a warning bell in my brain. I waved the tissue around without making contact.

"Mom, what if it burns?" I yelled, fighting the urge to imitate the raspy voice in *The Exorcist.* I hadn't seen it, but Tanya Tanner had. She said Linda Blair threw up black-eyed peas, then put a crucifix up her vagina and said, "Wow, that burns!"

"Oh, it's gonna burn," Mom called back from another room. "That's how you know it's killing germs."

I pitched the tissue into the toilet and reached for the jar of Vaseline.

A few hours later I was lying on my bed reading *Baby Island,* a book about two sisters who get shipwrecked with four infants on a tropical isle. They have to take care of them all by themselves, until they encounter a surly cockney chap with a missing toe living alone on the other side of the island. In exchange for cleaning his house, he gives them goat's milk to feed the babies. The younger, sportier sister, Jane, had a pet monkey named Prince Charley. If he were my monkey, I would have named him Taco. Or Bernard.

"Sarah!" I heard Mom call from out in the driveway.

I let the book drop and pretended to be asleep.

"Sarah, where are you?" Mom called again, drawing closer.

I jumped up and got the plastic puddle of dog vomit off my dresser. I bought it at the joke and magic shop in New Orleans with a five-dollar bill I found near the St. Anthony's poor box at church.

I lay back down on the bed and put it next to my open, fake-sleeping mouth. I stifled a shiver of delight. This was better than jumping out from behind corners and scaring people.

"Sarah, I've been calling you," Mom said, finally coming into my room. "I know you can hear m—OH MY GOD!"

I cracked open an eye. Mom was leaning over me. She didn't look worried, exactly; more tired of being worried.

"BOO!" I said. "HA-HA!"

"Dammit, Sarah," Mom said. "Do you want to give me a heart attack?"

"I'm fine. See?" I said, picking up the vomit and waggling it under her nose. "Jeez, relax. What's for dinner?"

"Your favorite," she said sarcastically.

Meatloaf.

Underground Railroad

"Mom, if you don't let me go to Camp Hickory Bluff, I'll just have to kill myself," I whined generically, wondering which personality I should click into for this, my latest entreaty.

Millicent Rose might work, but I had already used her once this week. Milly Rose was a studious, sensual girl, tailor-made to argue my case for wearing Mom's red nylon slip to bed. She wasn't exactly suicidal.

"Uh-huh," Mom said, hunched over her pregnant belly at the sewing machine.

I needed a fresh approach, someone with spunk. Jodie Foster was my go-to in these situations, but her career was flagging. She hadn't done anything of note since *Candleshoe*. There were more up-and-coming ascendants to the moxie throne. I'd seen Tatum O'Neal in *Paper Moon* and Kristy McNichol on *Battle of the Network Stars*. I admired Tatum's bootlegging swagger and the way Kristy crawled through a muddy trench in her satin jogging shorts. A tough yet pretty, camp-deserving persona could be cobbled together from Tatum and Kristy. That the two of them were currently starring in a movie all about summer camp, *Little Darlings*, was merely a fortunate coincidence.

"Mo-om," I said again, making my voice hoarse and perky.

Mom was sewing some flowery sundresses for me and my sisters. They had spaghetti straps and smocked bodices, the kind of dress that should be worn only by the flat-chested. Like Kristy-Tatum. Like me.

"Where is this camp again?" Mom said, snipping the thread with her teeth.

"Tuxedo," I said, rattling the brochure. "Tuxedo, North Carolina!"

"Smokey *Mountains?*" she asked, sitting up straight, flaring her nostrils like a hunting dog just picking up the scent trail.

"Blue Ridge Mountains," I said.

"This could really take care of your asthma once and for all!" Mom said. Her hands trembled as she read about the open-air cabins and nondenominational vespers.

Ecch. Kristy-Tatum was no asthmatic.

"I was thinking the same thing," I said wanly, slipping into Beth, the pale, gaunt, consumptive recluse.

It was a blatantly cheap *Little Women* rip-off, but Mom always fell for it. Standing behind her, I threw my arms around her neck and pressed my sunken cheek against hers.

"Oh, Marmee! Doesn't Camp Hickory Bluff look grand? I think . . . I could be . . . happy there."

Everyone was going to ritzy Camp Hickory Bluff, or at least everyone I cared about. Not the truly ultra-popular girls: they were truly out of my league. Their summers were taken up by hard-core cheerleading regimens that made their thighs curved, hard, and sleek as dolphins.

I'd heard about Hickory Bluff from the girls who had gotten their periods and were best friends with their moms. At school, they comprised a sort of mid-level clique; they had money but acted like it was no big deal. Later, in high school, they would become potheads and listen to Cat Stevens and Lou Reed songs about the wild side and the wild world.

Right now they were listening to Duran Duran's eponymous first album. They liked me because I could, on demand, write tales of domestic bliss starring them and whichever Duran Duran member they intended to marry. I would even include appendices of alternate floor plans for the sumptuous English castles they would share with Simon,

John, Nick, or Roger. (Nobody wanted Andy Taylor, the homuncular drummer; he was like the Terry Gilliam of the group.) Their lavish fantasy homes included a parlor, drawing room, sitting room, living room, family room, game room, rumpus room, gazebo, plenty of closet space for leather pants and ruffled poet shirts, and in the master bath, his and hers vanity tables with Clairol makeup mirrors for applying eyeliner and streaks of rouge. These girls weren't opposed to the idea of men wearing makeup as long as they played a mean synthesizer. And they were all going to this camp in North Carolina that sounded perfect. For me.

"Let's talk to your father," Mom said. "What's that whistling noise? You wheezing? I told you that Slow-Awphyllin stuff wouldn't work."

Slo-Phyllin was the latest drug I was taking for my asthma. It was a clear capsule filled with little white and purple balls of a time-release bronchodilator.

"Sure, I can swallow a cap-SOOL," I lied to Dr. Singh, jumping so fast at the chance for some free samples that I forgot to consider my overreactive gag reflex.

After several retching attempts to swallow the pill whole, I got inspired by that Contac commercial, the one where the capsule is broken open and its contents fly out. Every six to eight hours, I cracked open the pills over a plate and licked up the little balls. Within seconds of the balls dissolving on my tongue, my heart started to beat fast and I felt like bugs were driving tiny fire engines around inside my head. I was so disoriented I didn't notice whether Slo-Phyllin helped my asthma or not. In an hour or so, I would come down enough to find myself still wheezing, still coughing.

I'd been on the stuff for about six months.

"Dad, mightn't I please go to Camp Hickory Bluff?" I asked, earnestly clasping the deserving Beth's sickly, tapered fingers before me.

"Do what now?" Dad said.

He said this often, or sometimes just "Do what?" He was hard of hearing from working around noisy machinery all his life and probably just needed me to repeat what I'd said, but it always sounded like a challenge.

"Sarah needs to go to this camp in the Mountains," Mom pushed. "We think it would cure her asthma to get up above sea level."

"Cure her asthma?" Dad said. "Heh. What's it gonna do for her weird-shaped head?"

Pity. I remembered a time when he cared enough to give us clever, alliterative nicknames. I was the Cartwheel Kid, Becky was Pistol Pete, and Hannah was Toy Teeth. He'd really lost his touch.

Dad flat out refused. Mom and I reverted to ye olde bank swindle. Whenever the Mountains hung in the balance, her moral code reverted to that of Robin Hood. She would brave the potentially lethal consequences of stealing from the rich (Dad) to give to the poor (me). Mom's acute martyr syndrome beautifully complemented my chronic sense of entitlement to the finer things in life.

Mom found the checkbook at the bottom of Dad's leather camera bag. Once again, I went into Rebel Savings and Loan with a forged check. I signed Dad's name myself; I'd had lots of practice forging his signature on all my absence excuses at school.

"I can't cash this without proper I.D.," said the meddlesome assistant manager, his colorless eyes flickering with suspicion. "And something's telling me this ain't your checking account."

Dad must have tipped them off.

I don't know how she got it, but somehow Mom came up with money for the $200 that would reserve my spot by the fireside at Camp Hickory Bluff. We still needed another fifteen hundred. Mom contacted the camp and asked if they had a layaway plan. Amazingly, they offered no such option. It was cash on the barrelhead or nothing. Moreover, the fifteen hundred was just the base sticker price. There would be add-on costs, not the least of which would be the appropriate wardrobe.

Camp Hickory Bluff required a uniform of blue or white shorts. Mom went down to the Dollar Barn and bought me two pair of navy jogging shorts and, though it killed her, one pair of high-waisted white polyester athletic shorts. Mom hated polyester—she could smell it from a mile away. To hear her tell it, wearing polyester was like wearing a petri dish, its anaerobic fibers a rich nutritious medium for bacteria to bloom and grow, bloom and grow forever.

"I guess they'll be okay if you only wear them every other day, with two pairs of cotton undies beneath," Mom said, holding the shorts out to me like a turd on a stick. "But if it starts to itch down there, take 'em off and throw 'em away!"

With that rousing sendoff, I had no doubts the white shorts would itch immediately. Already I could feel a tingle creeping over my groin.

I was none too thrilled to own a pair of shorts that looked like they belonged to a nurse who did a little assistant basketball coaching on the side. In fact, my wardrobe for camp was keeping me awake at night. Blue and white dress code or not, I knew everyone would be wearing the latest Lacoste and Lilly Pulitzer fashions from D. H. Holmes and Maison Blanche and Godchaux's and that pricey boutique on Magazine Street, Perlis. Just the way people said Perlis—"PUH-liss"—sounded like their mouths were full of honey-roasted money.

"Don't worry, I'll get the rest somehow," Mom said, her eyes darting around the house, looking for items to pawn. "Wouldn't hurt you to chip in, either. Take all the sitting jobs you can."

✵ ✵ ✵

Back in January, I'd pledged to conduct a far-ranging anti-litter campaign I called Operation: Swept Away!

I had to present proof of twenty hours of community service in order to earn my 4-H merit pin for the year. As always, I dreamed big. Then I awoke and promptly forgot my dream. Sister Giles, retired from teaching but still in charge of St. Aldric's 4-H Club, reminded me of my pledge.

"How's your cleanup campaign coming along, S-s-s-s-s-sarah? Horrrrr-horrrrrrrAH," she asked, coughing wetly into a crumpled Kleenex.

"What? Oh, that. Just fine, Sister!" I sang.

Shit. It was already April. The annual 4-H Fair was in May, and those judges didn't fool around. They'd be wanting stone cold evidence.

"Mom, you were supposed to silk-screen fifty Operation: Swept Away! t-shirts, remember?" I complained. "Now it's too late!"

"Let's go over to Jonquil Street and clean up the park," Mom suggested. "I'll take pictures and you can show them to the judges."

Jonquil Street was the only street in the neighborhood with no jonquils. It was a short, scrubby little cul-de-sac with two houses separated by a vacant lot masquerading as a playground.

We were never allowed to play there. Rusty chains dangled swingless from a sagging A-frame. People dropped off their excess garbage there when they couldn't make it to the town dump. Judging from the pentagram spray-painted on the broken asphalt of the basketball court, a satanic cult met there. I imagined their first order of business would be to recruit a real go-getter like me. Thank God I was such a terrific Christian and immune to their satanic charms.

My sister Debbie held the garbage bag while I pitched in a few rusty cans, a strip of rubber tire, and a broken beer bottle. Mom snapped photos with Hannah's Kodak Disc camera. I started to *believe* in Operation: Swept Away!

"Watch out for tetanus." Mom warned, as though it were something that could be seen with the naked eye.

I was so busy orchestrating the cleanup, I didn't notice Willie Cox come out of his house. When I looked up, he was just there, assimilated out of thin air like a will-o'-the-wisp. I looked away fast.

Poor Willie was born with an arm that stopped short right after the elbow. Occasionally, the nub would sprout fingers, which would magically disappear a month or so later. I heard his mom took him to a doctor who burnt them off with sulfuric acid. Willie sometimes wore a brittle plastic prosthesis with an aluminum pincer at the end. Today he was au naturel, his stump hooked through the chain-link backstop, as if it were sniffing us. His eagle-eyed stare made me feel self-conscious, yanking me out of the moment. I knew he could tell we were just *pretending* to turn the park into a thriving suburban community center.

"Come on, let's blow this joint," I said to Mom. "You got the pictures?"

"I got at least one of you with that Mickey's Bigmouth bottle and about seven of Debbie with the garbage bag," she said. "Let's take one of you both with the Swept Away sign."

"Nah, no group pictures," I said, throwing the slack garbage bag into the back of our van. Two people did not a campaign make.

"Hi y'all!" Mrs. Cox waved from her porch. She walked over to us, wiping her wet hands on the wet front of her t-shirt. "You're Nancy, right? I've been meaning to ask if any of your girls would be interested in babysitting?"

I stole a swift glance at Willie. How could I watch a kid from whom I was always politely averting my eyes?

"What a coincidence!" Mom said, barreling on through. "Sarah needs a little extra money for this summer project she's working on."

She winked at me.

"I babysit," I said, picking up a beer bottle from the ground and peering into it.

I took the job, a regular Wednesday and Friday gig when Mr. and Mrs. Cox went to choir practice. Their house was plain but clean. Nothing special in the cupboards—no Cheetos, no Figurine bars—just stale, chewy graham crackers. The ones at the bottom of the package were still a little crisp.

Willie was six. As I predicted, it was impossible to keep my eyes politely averted. I had to remain vigilant at all times.

Willie was mad, and his stump was a formidable weapon.

If he couldn't open a box of crayons, or open the door to the back yard, or open a jar of pickles, he flung himself onto the ground, hollering. If I tried to help him, he charged at me, using the abbreviated limb like a tusk.

One night, I made the mistake of trying to help him snap some Legos together. With the precision of a professional wrestler, Willie whacked me in the forehead, clotheslined my neck, and stabbed me right in the pubic bone. I doubled over, paralyzed by a deep, electrical pain. I lay on the floor, pleading with my eyes. Willie merely laughed and swung the stump around casually, like a pistol.

Mr. and Mrs. Cox were so soft-spoken and serene, I felt too guilty to stop babysitting for them, even after Willie's assault on my reproductive area. He made me hate kids so much, I half-hoped he'd sterilized me for life.

I couldn't quit. I needed all the cash I could get for Camp Hickory Bluff. Nights, I lay awake, praying for a boulder to drop on the Cox home and for them to have willed any and all insurance monies to me, their beloved sitter. I wasn't wishing them dead, exactly, just more generous. Sort of.

In late spring, Willie got into some trouble at school. He broke into the tabernacle at St. Aldric's church. Father Adrian caught him hiding behind the altar, munching on Christ's body. Shortly afterward, the Coxes packed up and moved away.

By that time I had saved up about thirty dollars. It was such a piddling amount, it was hardly worth contributing to my Camp Hickory Bluff fund. The next time we went to Lakeside Mall in New Orleans, I felt perfectly justified in blowing it all on a cerulean blue Shetland wool sweater by Lacoste.

Ah, Lacoste! How I dreamed of owning a rainbow assortment of their puffy socks, folded under and over to center the alligator on my slim ankle. A sweater? Even better! I danced with it, snuggled up in it, laid it down gently, and made sweet tender love to it, and all before I even got it home.

"Here I am, breaking my back to send you to this damn camp, and you go buy an alligator sweater?" Mom carped as we unpacked the groceries we'd picked up on the way home from the mall. "It barely gets below fifty here and you're allergic to wool!"

"No, I'm not. Dr. Singh said I was allergic to every animal EXCEPT sheep, remember? Besides, I had to get it . . ." I trailed off, looking at the package of grayish pork chops in my hands. "It was reduced for quick sale."

Ever the bargain hunter, Mom couldn't argue with that. Secretly, I had to admit she was right. A wool sweater was a bit of a stretch for summer, even in the Mountains. I needed to get the most Lacoste bang for my buck.

Mom carefully clipped the threads holding the alligator to the sweater and pulled it off. Then she sewed the tiny reptile onto the left leg of the white polyester shorts.

"There you go," she said. "With the occasional spot treatment, you can wear these every day."

✵ ✵ ✵

Mom and I revealed our plan to my sisters one by one. Everyone seemed delighted to be in hush-hush cahoots against my father. They had all lived in fear for their own hides at one time or another. Seeing one of us—*me!*—nearing the verge of escape kept them from crying, "No fair!"

I was happy to be the sole beneficiary of our intricate scheme.

Why not me? I could get used to that. I *needed* to get used to it. I deserved to. It sure beat *Why me?*

Through some sort of elaborate, last-minute cash-juggling scheme, Mom got the rest of the money and wired it two weeks before I was to go to Hickory Bluff. I was learning invaluable lessons about personal finance.

"If our timing's right, this withdrawal won't appear on his bank statement 'til after you're gone," she murmured, combing through Dad's file cabinet for supporting documentation.

We borrowed a steamer trunk from her English friend Phoebe. Mom was enamored of Phoebe because she had watched her own C-section in a mirror at a British hospital. Mom was very much in favor of watching babies emerge from their mothers. In these last months of her pregnancy, she was constantly referring us to *A Child Is Born,* a book with photos of fat hippie ladies grimacing and grunting as wrinkled babyheads stuck out of their vaginas.

"Isn't it a miracle?" Mom would sigh.

"Um, yeah," I'd swallow, wondering if long ratty hair and mailbag boobs were prerequisites for having a baby.

The plane tickets were bought and paid for. There was only one thing left to do: come up with an excuse for my prolonged absence in case Dad noticed. We decided to pretend I was at Camp Olde Cypress, a much cheaper operation ten minutes away from our house.

"We should go check it out. He might ask questions," Mom said. "Olde Cypress is in session next week."

"I catch your drift," I said, remembering the full-immersion method I used to "become" Sister Titus.

This role would be more challenging. I would be playing the part of a girl who wanted nothing more than to stay within a few miles of her own backyard. I would need a little emotional recall and fake memories strong enough to give a convincing performance and counteract my stage fright, should my father and I have a scene.

At the ramshackle gates of Camp Olde Cypress, Mom and I were met by Monique, a junior counselor. She seemed less than thrilled at having to show us around.

"I'm supposed to be over there at JC council," she said, pointing to a gang of preteens smoking out on a patch of gravel next to a greasy, half-pitched tent. "So what, you coming here or not?"

"Well, I'm really going to another camp—" I began.

"You come here, I'll mess you up, you try to mess with me," she said.

"Now—Monique, is it?" Mom said, putting her arm around the girl's shoulders. "If Sarah messed with you, *I'd* mess *her* up."

"Sheeeeeeee-yut," Monique said, looking at my mother the way skeptical people in commercials look at something they think will not be delicious to eat, then when they try a bite, they realize how delicious it is after all.

Situated on a congealed, chartreuse pond, Camp Olde Cypress was a sweltering swamp of inactivity. No crisp mountain air for these campers. The few we saw that day lounged in the scant shade, scratching at their ringworm and lethargically swatting at encephalitis-carrying mosquitoes. Nothing—no laughter, no song, not even the sound of a lanyard being woven—interrupted the cicadas' "wee-ewwww, wee-ewww, wew-wew-wew-wew."

"Thank you much, Monique," Mom said.

We walked fast toward our car.

"That's right, you better run," Monique said lackadaisically, out of vinegar.

Speeding away from that furnace of indolence, Mom and I felt justified in our commitment to my living beyond our means.

"Did you see that tent?" I snorted. "It looked like a dirty napkin at a fish fry."

"I didn't see a single cypress tree," Mom sniffed. "Olde or otherwise."

✵ ✵ ✵

I was feeling pretty spiffy when I marched onto the Piedmont Airlines flight to North Carolina. Settling into my seat, I introduced myself to the girl next to me. She was wearing pristine Nike court shoes.

"My name's Sarah," I said, pulling my nondescript but new t-shirt down over my old mauve denim wrap skirt. "Are you going to Camp Hickory Bluff?"

"First year, huh?" the girl said. "This is my third year."

Seniority trumped all. Damn. There was no way to fake seniority. I should have worn those white polyester alligator shorts on the plane.

Fooling my father was one thing, but there was no way to conceal my indigence from the ruthless future debutantes at Camp Hickory Bluff. I had the two blue shorts, one pair of white, and five plain cotton t-shirts in a couple of boring colors. Camp regulations dictated that we wear blue shorts every night and the white on Sundays, so I wasn't going to get much mileage out of that gator after all.

The only time we weren't in uniform was during our morning classes. If I signed up for strictly aquatic activities, I could practically live in my classic racerback Speedo. It was from Goodwill, faded and pilled at the crotch, but I could blame all that wear on a long, imaginary Easter vacation in Cancun.

To deflect attention from my wardrobe, I butched out in canoeing class, learning all the strokes in record time. Every morning I got up at six AM for Ice Breakers, an exercise in machisma that found me voluntarily jumping into a freezing-cold lake with all the short-haired Hickory Bluff campers, while the rest of the girls slept in. I crafted friendship bracelets for the most casual of acquaintances. Down at the camp pool, I competed ruthlessly in every evening's greased watermelon contest. Funny, nobody hoisted me onto their shoulders and paraded around camp in exultation.

Starved for admiring companionship, I looked to the Mountains. Perhaps they would reinvigorate me. The Mountains stood cold and distant, ever taciturn. So, I embarked on a persecution campaign against the only girl at Hickory Bluff who might possibly be as poor as I was.

Believe me, I had to do some research just to be on the safe side. Her name was Missy Gatlin—who's to say she wasn't descended from the pioneer founders of that folksy enclave in Tennessee? The fact that she had only two outfits could very well be an indication of eccentric aristocratic inbreeding.

I watched Missy closely the day we took a field trip to the Biltmore Mansion in nearby Asheville. She seemed uncomfortable amid the splendor of its 132 rooms, spending most of her time away from the crowd, wandering beneath the majestic pines. At lunch, an egg salad sandwich in hand, Missy used her pinky to push her glasses back up on her nose. That pinky move . . . pret-ty regal . . . then she shoved the whole sandwich into her mouth, gave a gulp and burped, then chewed and swallowed again.

Whoa: raised in a barn, or too rich to care?

"It's quite possible I am a distant relation to these here Vanderbilts," I proclaimed within Missy's earshot.

Not a peep out of her. Proceed with persecution as planned.

Missy was a junior counselor and was often assigned to babysit our cabin during rest hour, when the senior counselors were off on a smoke break. She was only a year older than we were. Missy was like a clerk at a record store, bossing us around from her barely elevated bully pulpit.

"You're being entirely too noisy in there," Missy would say from the porch of our cabin.

"No," I answered back one day, perfectly mimicking her voice, "YOU'RE being entirely too smelly out there."

Imitation is the sincerest form of pandering. The laughter of my cabin mates felt like a warm caress. Clever cruelty might just be my ticket to instant popularity.

"Who said that?" Missy said, sticking her head in our cabin door. "I'll be waiting outside, expecting an answer."

She closed the door.

"Jedi you are," I croaked, wielding my uncanny Yoda impersonation like a light saber.

The girls cheered me on.

"I'm writing you all up!" Missy snapped.

"Hmm," I croaked again. "Jabba you are."

That day, during free time after rest hour, my cabin mates fought over who got to share their contraband Hershey bars with me. Courtney Larpenter volunteered to show me how to insert a tampon.

"Don't do it the way it says to on the package," said Courtney, lying on the cabin floor with her knees up. "Sticking it in hurts like hell—ow!—but once you get it through that wall of muscle, you can't even tell it's in there, I swear. See? Now nobody has to see that nasty mess."

Over the next couple of weeks, girls did things like trip Missy in the dining hall or hide her glasses or pull Pixie Tricks on her. Named after Hickory Bluff's cute elfin mascot, a Pixie Trick could be any standard camp prank, like crushed cookies in a sleeping person's ear or shaving cream in someone's shoes.

Sunday night we had vespers, a song and lite prayer session held in a clearing in the woods. In between Joni Mitchell's "Both Sides, Now" and Kenny Loggins's "House at Pooh Corner," the insults being hissed in Missy's direction built to a crescendo malo.

"Gatlin," someone said. "More like *Fatlin.*"

Fatlin–Fatlin–Fatlin–Fatlin . . .

I never meant for it to go this far. Things were spinning out of control. I had to stop it before the girls chased Missy into the woods and raped her like the cool kids did to the nerdy girl in that movie *Last Summer.* A young Barbara Hershey played the alpha girl who directed the raping. I liked to imagine myself looking as taut and tan as Barbara had in her string bikini, but ultimately, I couldn't have that sort of thing on my conscience.

"May I be excused?" Missy asked the head counselor.

She stood up, lost her balance, and nearly fell into the campfire. When she turned to leave, everyone saw it.

Missy had gotten her period. In her Sunday white shorts. The vultures descended, picking the last shreds of meat from Missy's bones. There was a lot of meat on those bones.

I thought it was just plain sad. For me, not for Missy. So she might be poorer than I was—so what?

She had gotten her *period.*

So what if I could rag on a fattie in the voice of a wizened green midget creature from a galaxy far, far away? I had influenced people, but had I won any friends?

I abandoned my leadership post in the Fatlin Brigade. Instead, I spent extra hours in the Craft Barn, making a gilded leaf brooch for Hannah and a googly-eyed peanut on a mossy rock saying "Hang in There" for Debbie. Fueled by guilt and something else—regret, perhaps—I logged twenty hours cutting, sanding, and assembling pieces of birch into a musical jewelry box for Mom.

"Wood is good," said Ben, the woodworking instructor, handing me some tiny nails. It was something Ben said often, in a hypnotic voice, like a mantra.

I glued in the Swiss musical movement I'd specially ordered. It played "Lara's Theme" from *Dr. Zhivago,* one of Mom's favorite instrumentals.

In my bunk after lights out, I quietly kneaded and punched my belly to hurry things along.

In her letters from home, Mom kept me updated on how our lie was holding up. Dad continued to believe that I adored Camp Olde Cypress so much I just couldn't tear myself away for a visit at home.

Father's Day was coming up. All the other girls were making presents for their daddies. I should make one for my daddy, too, right? By calling him "my daddy," I thought I could recapture some of the magic that had never existed. Or had it? We had a special closeness, didn't we? I remembered our fishing trips, only the good parts: the soda, the sandwiches, and my ability to catch giant redfish and a wicked sunburn and chug Tabasco. All that had to count for something.

My friendship bracelet endeavor had pretty much petered out. I decided to make my daddy a sand candle.

I made a hole in some sand, dropped in a wick, and filled it with molten wax. It looked like an overgrown GooGoo Cluster, but the wax was a nice hunter green. After the wax had cooled, I pulled it out by the wick and etched a Father's Day message on the top with a straight pin: "Please don't be mad. Love from North Carolina!"

"This candle should take care of everything," I said to myself, pursing my lips and cocking my head a quarter turn.

"That looks nice," a voice said from behind me.

"Thanks, Missy," I said.

I stayed late at the Craft Barn that night, making Missy a red-and-white friendship bracelet. I'd give it to her anonymously, of course, but it was the thought that counted. Not telling anyone about it would make it an act of charity, more Christian.

When Dad's candle arrived back home, Mom put the package on the foot of his bed. They didn't sleep together anymore. Dad got home a few hours later and went to his room to take off his pants. I imagine Mom, pregnant, and Becky and Hannah and Debbie all huddled in the kitchen, holding their breath, listening to him rustle through the packing material.

Dad came out of his bedroom with the candle, walked over to his desk, and set it down among all the other gifts we'd made for him over the years: orange juice can pencil cups, misshapen ceramic ashtrays, plaster casts of our open hands. He went into the kitchen, walked past Mom and my sisters, got a bag of sunflower seeds out of the cabinet, and took it to his recliner in front of the television. He didn't move until they were all sucked and seeded and his tongue was white with salt. Then he got up and went to bed.

I know this because Mom wrote it to me in a letter. As I lay on my bunk reading it, with each sentence I felt like I was plunging down through another floor in a tall building.

✵ ✵ ✵

Two weeks later, I arrived home from camp full of 100 percent pure fear, coated with a crispy candy shell of silly stories and sassy song.

Spending five weeks away had dulled the memory of my father's totalitarian regime. He'd probably haul off and slug me for sneaking away, but the physical pain wouldn't be real until impact. After impact, it would fade.

I was more afraid of my mother. The girls at camp had brainwashed me into shaving my legs, something Mom had strictly forbidden me to do until I was fifteen. I had come to savor their sleek feel and couldn't wait to introduce Becky to the wonders of razor depilation and daily leg oiling.

"You can go wipe that lip gloss off right now," Mom said when I stepped off the plane in New Orleans.

Right away on the ride home, she noticed my unnaturally smooth and shiny legs.

"I just don't know how you could go behind my back and do that, how you could defile the pure gifts of nature—"

"Leg hair is a gift of nature?" I asked, incredulous.

"—right in the *heart of nature,* after all I went through to send you to that camp in good faith, and this is how you repay me."

I started singing my favorite new song.

Ask mom and dad for lotsa dough
Cause you really want to go
To Camp Hickory Bluff
Oh, you'll never get enough.

"Your father will back me up on this," said Mom, louder. "He's none too happy about your waltzing off to that camp in the first place, I can tell you that."

"And the dough won't matter," I sang at the top of my lungs, "When you tell 'em 'bout the fun / You had swimming in the lake and playing in the sun."

I pictured Missy Gatlin's parents wearing colonial garb, welcoming her back home. Missy and her mother would sit down at the butter churn and have a heart-to-heart.

Their house looked like Monticello.

Next year, I'd have Mom sew the alligator onto a pair of 100 percent cotton shorts.

A nice poplin, or maybe seersucker.

The Center Cannot Hold

"Hurry up, get up, get up, hurry," Mom said.

With one arm she dragged me out of sleep into a standing position, balancing a nodding Debbie on the opposite hip. Next to the silhouette of Mom's eight months' pregnant stomach was Hannah, whimpering and rubbing her eyes.

"Come on, get dressed."

The orange numbers on the digital clock said 12:37. I tucked my nightgown into my jean shorts and yanked on a pair of rain boots, whatever I groped out of the heap in the corner of my bedroom. I heard crashing and banging at the other end of the house. It sounded like Godzilla attacking Tokyo.

"You're not goING GOD DAMN ANYWHERE."

Mom hustled us out to the car. Becky and Hannah ducked into the backseat. This was no time to argue over who would sit shotgun.

"GET OUT OF THAT CAR," Dad said, barreling out the back door.

He grabbed Debbie, who was closest to him. She flopped like a feather pillow. It reminded me of the scene from that PBS nature show, when the killer whale roars up onto the beach and snaps up the seal pup. He tossed Debbie under his arm and started limping back to the house.

I ran after them, inserting my body between them and the door.

"Put her down!" I said. "Leave her alone!"

Dad opened his free hand. I didn't back down. I was pretty sure I could take him. If Pippi Longstocking could lift a horse over her head, I should be able to whup my dad.

I could at least outrun him, if it came to that.

Debbie slid down the side of his body, landing in a soft heap. I clasped her hand and we backed away to the car. I tucked her into the rear seat.

Ha! Front seat, suckers!

Since I got home from Camp Hickory Bluff, this sort of thing was happening more and more. Usually Mom just drove around the Winn-Dixie parking lot for a few hours, until she figured Dad had cooled down and it was safe to go home.

Tonight, in an inexplicably freak instance, Mom had sixty dollars in her purse. We just kept driving until we got to my Aunt Carol's in Kansas City. She still lived in the same stately, grand old home she'd always lived in.

Cramped in the Gremlin on the interminable drive, Becky and I tortured Hannah by sticking wads of chewed gum on her skin when she wasn't looking. I hated buttons and mayonnaise; Hannah hated gum.

"What a terrific surprise," Aunt Carol said when we appeared half-clothed and hungry on her doorstep. She took us in, fed us Apple Jacks and Taco Bell Enchiritos, marinated artichokes and salad with chunks of real blue cheese in it: not just whatever we wanted, but things we never even knew we wanted.

Our first night there, Mom went into my aunt and uncle's bedroom to call Dad and tell him where we were. She emerged shaky and red-eyed.

"I can't go back," she said to Aunt Carol.

"You have to go back," my aunt said.

That week Aunt Carol hosted an impromptu family reunion of all our relatives who still lived in the area.

"Becky's getting hooters," Uncle Tim observed as Becky rose from the Jacuzzi in a wet t-shirt and underwear.

Unlike me, she hadn't thought to grab her swimsuit as we escaped our father.

"C'mere, gimme a kiss!" Uncle Joe said, drawing me into his arms, his mouth full of chewed-up ham sandwich.

I veered my head to the side, aiming for a less intimate embrace, but Uncle Joe wouldn't hear of it.

"Nuh-uh! None of that cold stuff!" he smiled, a frappé of mayonnaise and ham bubbling between his teeth. "Right here! On the lips!"

I slid into the Jacuzzi. Uncle Joe's kids, my ten-year-old twin cousins Rider and Ranger, were whipping the waters into a froth with their scrawny bodies.

"Can you believe we're related to these people?" Ranger gleefully asked me.

"Yeah, these people are *rich*!" Rider said, trying to dunk my head under the bubbling water.

"Excuse me," I said, rising sedately from the hot tub.

I walked over to a chafing dish of shrimp dip, looking to hobnob with someone more my type.

"Come with me for a sec, Sarah," Aunt Carol said.

She led me down the hall into the darkened formal dining room.

"Sarah," she said, cupping my shoulders. "You're like me, I can tell."

"Yes?" I said, not knowing what she was getting at.

"You're the strong one in your family."

I flexed my biceps.

"It's not easy to be the strong one," she continued. "If you ever need to talk to someone, you come to me."

My throat hurt.

"Aunt Carol," I said.

"Yes, hon?" she said.

I swallowed.

"Can I open another can of Pringles?"

We spent a week at Aunt Carol's. I wanted to move in forever.

✵ ✵ ✵

"You better stay home from work today," Mom said to Dad at breakfast one morning that October. "I feel like the baby's going to come today."

We still had no idea whether Mom was having a girl or a boy. She was forty years old, but they didn't do sonograms back then. She probably wouldn't have had one anyway. Mom was doing everything natural this time: no cigarettes, coffee, booze, uppers, or downers. She planned to have the baby without anesthesia and breastfeed it, too. As usual, she would use cloth diapers instead of disposables because they were more eco-friendly, and also because, as she said: "I wouldn't want a piece of plastic taped to my butt. Would you?"

When we got home from school, there was a note from Dad telling us they were at the hospital. We waited around. Finally, at six-thirty, Dad called.

"Your mother's fine; the baby's fine," he said, sounding genuinely elated, almost woozy with glee. "It's a boy!"

A boy!!

Mom wanted to name him Stephen Francis and Dad wanted to name him Patrick Allen. They compromised and brought home little Michael Lawrence.

"Oh my god, Mom, don't you just love him?" I cried, standing over him while she changed his sopping cloth diaper. My heart felt like a water balloon being filled up with a faucet on full blast.

"No," Mom said.

"*No?!*" I screeched.

"Well, I don't know him yet," she smiled. "When I know him, I'll probably love him."

"You're a terrible mother!" I grimaced.

She had given birth to Michael the way she wanted, naturally. No anesthesia, nothing. I didn't see what the big deal was.

"It couldn't have been that hard—you already had four of us," I said. "Isn't your vagina like, totally stretched out by now? He probably just slid right out, huh?"

"You were my easiest baby, Sarah," Mom said, down on her knees, swishing a poo-filled cloth diaper around in the toilet. "You were born half an hour after I got to the hospital."

"Becky must've carved you out pretty good," I said.

✲ ✲ ✲

One day that fall, on the playground at recess, I saw a fourth-grader wearing a bandana on her head, Aunt Jemima–style. As an eighth-grader, I normally didn't bother with nine-year-old peons, but I couldn't resist.

"Hey Mammy!" I called. "Oh Mammy, how I love you! My dear ol' Mammy, my rag-time ga-aaaaallllll!"

"What are you DOING?" Michelle Banks said, staring and glaring at me and grabbing my arm.

"Look! She's like an ol' mammy!" I laughed, pointing at the little girl, who was also staring and glaring at me. "Love ya mammmmm-MAY!"

I wasn't being racist. I was imitating Bugs Bunny imitating Al Jolson, with a little bit of that "Hello my baby, hello my honey" Warner Brothers frog thrown in. Anyway, the little girl was white. She was Darren Johnson's sister—oh my *God*—the one who had leukemia last year. I remembered that right when Michelle said, "That's the Cancer Girl, stupid!"

I choked and sobbed and ran away to the bathroom, where my chokes and sobs echoed off the cool marble walls. I cowered in a stall and scooped water out of the toilet to wipe my face, because that's all I deserved: toilet water.

"Sarah?" It was Michelle Banks.

"I feel terrible," I said, coming out of the stall.

"Don't," Michelle said. "Cancer Girl's going back in the hospital for a month tomorrow. She says she doesn't want to fight you 'til she gets back."

✲ ✲ ✲

Vic and Ruth were staying with us over the Thanksgiving holidays. I'm not sure how Dad knew them; they were an unmarried, childless couple who just sort of appeared in our lives.

I liked Ruth because she had long, curled eyelashes and looked tough in her tank tops. Plus she knew how to scuba dive.

"What do you do if a shark comes up to you?" I shivered.

"Punch 'em in the nose," Ruth said, stretching her long thin arms over her head, displaying fine downy puffs of hair in her armpits.

Vic had a wide, doughy face and permed brown hair. He breathed noisily through his nose, sauntering around our house in a long striped Moroccan nightgown.

On Thanksgiving Eve, Vic was sitting in Dad's recliner, his right ankle slung over the opposite knee. I was sitting on the floor up close to the TV watching the *Love Boat.*

"Hell, that's the fella from *Mary Tyler Moore*!" Vic wheezed, tapping his pack of Pall Malls against the end table.

"He's called Captain Stubing now," I said, turning around with intent to lecture.

Vic exhaled a squeaky double-stream of smoke through his nose like a mad bull. Beneath the striped Bedouin tent of his man-gown lay a tangled mass of flesh and hair that looked like one of those hedgehog foot wiper/doorstop thingies.

I turned back to the TV, flustered. Captain Stubing bowed his bald head and smiled condescendingly at a pushy passenger, a real Mrs. Van der Snoot type.

"Anything else on?" Vic asked.

I switched the channels. Bees were swarming over a Volkswagen bug parked in the middle of the Superdome.

"*Killer Bees!* All right-y. Now this is more my speed," Vic said. "Killer bees're on their way up from Africa as we speak. Probably make it here by the end of next year."

I got down on my belly and wriggled my way across the floor and down the hall to my sisters' bedrooms. As per my whispered instructions, they paraded one by one through the family room, ever so subtly, to take a look at Vic for themselves.

✲ ✲ ✲

Thanksgiving morning I had to go to the emergency room because blood was dripping out of my ear.

"I don't know how it happened," I told Dad. "I was just gently scratching the outside, when *ploof*."

I tightly palmed the bobby pin smeared with ear wax. If we'd been a normal family with Q-tips in a glass canister on the bathroom counter, this wouldn't have happened. Now I had a medicated wick of cotton hanging out of my ear.

"I suppose you and Becky could have a touch of wine with your dinner today," Mom said, putting her traditional Thanksgiving platter of olives, carrots, and celery onto the dining room table. "It would go well with the cheese soufflé."

Mom's gourmet streak still flared up every once in a while, usually around the holidays. This Thanksgiving she roasted a goose instead of the usual turkey. The soufflé was her first, something she kept calling an "appetizer."

Appetizer: the word made me think of a ride that made you want to eat. *Please keep your arms and legs inside the Appetizer at all times.*

I could use a few hours on the Appetizer: my asthma medication kept me too jumpy to eat. At thirteen years old, I weighed about fifty pounds.

The soufflé was not appetizing. It tasted like slimy cheese air and was no better washed down with wine, and even worse coming back up onto my plate.

"You made yourself do that," Dad said, embarrassed in front of Vic and Ruth. "Clean your plate."

I looked at the goose leg on my plate in its puddle of grease, with a side of regurgitated soufflé. I swallowed hard.

"You mean I have to eat the vomit?" I squeaked.

"Finish your dinner," Dad said.

The conversation around the table slowly resumed and Dad forgot about checking up on me. The thrown-up soufflé congealed on my plate. I ate an olive and gagged down the rest of my wine. I felt a warm band tightening across my chest, like a heating pad crossed with a blood pressure cuff. I inhaled and exhaled. It wasn't an asthma attack. In fact, my lungs let go, hanging soft and slack inside my ribs for the first time in months. It was like the best hug I ever had.

"Are you going to drink that?" I asked Becky, nodding toward her glass.

"Uch, no, it tastes disgusting," Becky said.

So I drank hers, too. The warm band grew tighter.

"I'll clear the table!" I announced, when everyone had finally stopped eating and pushed back their chairs.

"Don't forget to sweep up," Dad said.

"No problemo," I said, shooing everyone from the dining room. "Go on, go relax! I'll take care of the kitchen."

I rushed back to the table. Looking over my shoulder, I began stacking the dirty plates. When everyone was truly gone, I clanged a knife against a plate with one hand, while using the other to grab every wineglass, every Lowenbrau bottle, every short, old-fashioned tumbler on the table and swig their remains.

When I awoke, my mother and Becky were sitting next to me on the bed.

"Jesus, Sarah, you scared the shit out of me," Mom said, shaking her hands out from the "wring" position and stalking out of my bedroom.

"What happened?" I asked Becky, propping myself up on my elbows.

"You chased Hannah and Debbie around, whipping them with an extension cord," Becky said, looking at me warily. "Then you climbed up high in the mimosa tree out front and wouldn't come down. You hung by your knees and ate your own booger and asked all of us to eat your boogers and then you said you'd eat all of our boogers. We couldn't get close to you because you were still whipping the extension cord around."

What a gyp. I didn't remember a single bit of it.

"Tell me again!" I said.

* * *

Back in the 1930s, Grandma Vivian and Grandpa Harold got married in a big band fever. They went swing dancing every night until Grandma got pregnant with my father. The fever turned ice cold. They divorced.

A single working mother, Grandma raised my dad alone, with some help from her mother, my Great-Grandma Gertie. Gertie worked in a bakery and taught my dad how to ice cupcakes extra swirly. She had come over from Sweden as a girl. She was dour and terse. Sometimes she came out of the bathroom with her upper lip all red and shiny and little daubs of shaving cream in her ears.

Grandma Vivian worked for years at Swift Meats. She brought us cases of bacon and, later, Sizzlean.

"Move over bacon, now there's something meatier!" I'd sing, lustily biting into striped strips of meat that resembled cross-sectional diagrams of the earth's strata and, coincidentally, tasted like volcanic ash.

When we lived in Kansas City, getting a solo weekend at Grandma Vivian's apartment felt special. She'd take me to the grocery store and let me pick out a snack. Invariably, I picked a six-pack of tiny wax bottles filled with colored sugar water.

Grandma let me dig through her giant chest of costume jewelry and puff on as much Coty face powder as I wanted. I would lie on her tufted teal carpet watching *Kojak* reruns until 3 o'clock in the morning, enveloped in clouds of smoke from her ever-lit True. We didn't talk much. Grandma drank scotch and water and I drank my sugar juice. Sometimes she'd doze off, snorting softly, her hand at rest inside the neckline of her blouse.

After we moved south and she retired, Grandma Vivian became a globe-trotting bon vivant. Between trips, she'd come visit us in Louisiana, bearing gilt-edged playing cards from the Liberace Museum in Las Vegas or Lucite watches from Hong Kong. I loved my plastic watch. Through a clear panel on the back, you could see the gears moving.

Grandma Vivian hadn't been to visit in a while. She and Mom didn't get along. On her last visit, when we still lived in Houma, they'd had a huge argument.

"Himph, when did beer become a breakfast item?" Mom asked her.

"If you keep picking on me, I'm leaving!" Grandma said.

"Go ahead, you big baby!" Mom said.

"I hope the next time I see you is at my funeral!" Grandma said, flinging her matching set of seafoam green luggage into the taxi and leaving for the airport three days early.

Next thing I knew, Grandma was a bleary-eyed semi-vegetable down the highway at Sunset Plantation Nursing Home. Her normally made-up face puffed and sagged like Bert Lahr's.

"What happened to Grandma?" I asked.

Mom offered only vague explanations.

"One night her friend Lena found your grandma passed out cold on the floor of the cruise ship ladies room," she would say, or sometimes: "She'd forgotten to eat and was living off scotch and menthol cigarettes and Lena found her knocked out cold on the floor of the bathroom at the Golden Nugget casino."

The particulars varied, but she was always passed out, always cold, always on a public restroom floor, always found by trusty Lena. Grandma Vivian's party was over.

I was quite familiar with the residents of Sunset Plantation, having frequently entertained them with St. Aldric's 4-H Club. Over the years, I'd improvised countless sassy shimmies to every tune from the jazziest album in my parents' music collection, Herb Alpert's *Whipped Cream and Other Delights.*

My last number had been a pulsating, Fosse-esque performance to the song "Green Peppers." This tune worked well with fringe. Thanks to my asthma medication, I still fit into the same orange-sequined dance costume from Goodwill.

On a makeshift stage in the O.T. (occupational therapy) Room, I shook my fringe and waved jazz hands for an audience of seniors. Most were slumped in their wheelchairs, drooling and mumbling. One exception was Regina, a towering pole of a woman who paced the room, stage and all, swiftly and purposefully. Her left eye was covered by a cloudy membrane, like a patch made from thin skin, with a thick vein running through it. Regina was nimble and alert. I had to work around her.

"Regina, sit down!" the more sedentary old ladies cried, craning their necks to see me rubbing myself against a table.

A man with an extraordinarily elongated head began to emit loud, rhythmic flatulence.

"Jake, git on," chorused the wheelchair ladies. "You're smellin' up the room!"

After our bows, Sister Giles nudged us into the urine-soaked crowd.

"Bid adieu to your elders," she said. "Use your Fuh-fuh-fuh-french."

I was cornered by Kitty, a sweet little old lady with a wispy topknot on her head tied with a pink paper ribbon. She had some sort of fluid-filled blister on her lip. It looked like a small bubble-gum bubble stuck onto the corner of her mouth.

"Au revoir!" I said extra-cheerily, patting Kitty's topknot and trying to get away with a loose hug.

"Ici, ici, cher!" she said, pointing to her bubble-lip.

I closed my eyes, bent over, and puckered gently, praying it wouldn't pop.

When Grandma Vivian moved into Sunset Plantation, we went to visit her at least once a week. Its distinct perfume of urine, Lysol, bouillon cubes, aging flesh, and Royall Lime hair tonic became less and less of an affront, as Grandma became less and less recognizable.

Her hair, which had always been coiffed into tight shiny blonde configurations, thinned and grayed within a month of being at Sunset Plantation. We came in for our weekly visit and found two black nurses giving Grandma corn rows.

"Oh," Mom said. "That looks . . . nice."

The nurses got up and bustled around the room, leaving Grandma's hair half-braided and half not.

"You gonna change the twinkie in #4?" one of the nurses said.

"What's it look like I'm doing now?" said another, lifting my grandmother by the ankles.

"What's a twinkie?" Mom asked.

Sometimes, her tactlessness could handily satisfy my curiosity.

"Twinkie's a poop diaper," the first nurse said, laughing.

"Twinkie? Why do you call it that?" Mom pushed further, her eyes sparkling. *Enough!* Phantom Me leapt toward her, arm outstretched to slap a hand over her mouth before she could say what she said, which was: "Because it gets *cream filled?*"

The nurses exchanged glances that said, "White lady crazy as shit."

"Your grandchirren are here!" they screamed into Grandma's face before leaving the room.

My sisters and I sat at Grandma's bedside, babbling uncomfortably in spurts and then falling silent.

"I don't like going to see Grandma," I complained in the car on the way home. "I don't have anything to say."

I felt deprived when I was less than my usually scintillating self.

"Just talk about your life," Mom said.

That sounded like the rudest thing ever.

Hey, Grandma! Guess what? Today I jumped and played and sang and smiled! Sorry you have to lie here staring into space all the time with a cream-filled Twinkie taped to your butt.

In the end it didn't matter what I said to her. Her former chattiness was gone, her wanderlust neutered, her will to argue with my mother and slug down booze and chain-smoke: all sapped. The only two words she ever said now were "Not necessarily," and she was dropping syllables as the days went by.

"Grandma, did you have a good dinner?"

"Not necessarily."

"Grandma, do you want to go outside?"

"Not ne'ssarily."

"Grandma, we're leaving. We'll see you next week."

"Not n'arily."

"Narly."

"Nnnn."

❋ ❋ ❋

Bip looked just like the Jesus from the TV miniseries *Jesus of Nazareth:* shaggy hair, sunken cheeks, scraggly beard, and clear ocean-colored eyes. He walked like Jesus, too: the way Jesus would have walked uphill, bearing the weight of a giant wooden cross. His spine was curved forward, and a congenital birth defect made his right knee point in toward his left, buckling his legs into a "K."

On Sundays, he took up the collection at St. Aldric's, always lagging about ten or twenty pews behind the other three collectors. He took so long, we'd reach the seldom-sung third and fourth verses of hymns: the poorly written, arrhythmic, scary verses about Satan and Hell.

One Sunday, the congregation was stumbling through the obscure, penultimate verse of the already torturously atonal "O Come, O Come Emmanuel." Craning their necks to see if Bip was done with his assigned quadrant, everyone's voices thinned to a shrill, noncommittal octave.

Today, Bip was really behind. Father Adrian launched into a second song:

Sing of Mary, pure and lowly
Virgin Mother, undefiled!

The tinkling of coins signaled that Bip had dumped the collection money into the wicker hamper at the rear of the church. To speed things along, Father Adrian cut everyone off mid-hymn:

And that's why we love you, Lor-or-ord!

After Mass, Mom and Bip stood in the doorway of the baptismal font, chatting. Dad still never went to church, so she was free to flirt.

My sisters and I waited on the steps of the church, impatient and hungry. Michael squirmed in my arms, his wet diaper alone weighing at least ten pounds.

"Oh, Bip," Mom said. "You're just filled with the Spirit."

"See you next Sunday, Miss Nancy," Bip grinned, flashing his cylindrical top teeth.

He shuffled down the church steps and mounted his giant tricycle, a terrific mode of transport. Perhaps this was a relationship with some promise.

"Hey, Mom, is Bip your boyfriend?" I asked in the van on the way home.

"Sarah, no!" she said, driving with one hand while she nursed Michael on her lap. She blushed and giggled and wouldn't answer the question.

By the following Sunday it had become part of the vernacular, as long as we were out of Dad's earshot.

"Gotta go to church and see my boyfriend," Mom said.

"Ask him if I can drive his tricycle," I begged, with Hannah and Debbie fighting over who would get to ride in the rear basket.

My mother, who was perfectly capable of asking the Sunset Plantation staff embarrassing questions about poo and pee, was suddenly far too shy to ask Bip if I could ride his trike. Would I never feel the caress of admiring glances as I rode off into the sunset on that lovely three-wheeler, with its hand pedals, flapping orange flag, and dangling coon tail?

However precariously, my parents were still married. Any potential romance between Bip and my mother died on the vine. I stayed on polite terms with Bip, just in case.

"Sure wish I had a trike like that!" I said whenever I saw him, hoping he'd make me an offer.

✵ ✵ ✵

At St. Aldric's Spring Fair, I played Wheel of Fortune and won a bottle of red wine.

"Wooooo!" I said, thinking, *Liquor time!*

"You need your momma or deddy or an ADD-ult to come sign for this," said Mrs. Perkins, the school secretary. She always ran the wine booth, probably because she was too dumb to drink it all up.

A few feet away at the white elephant booth, Bip was hanging out with T-Paul, the deaf-mute school janitor.

"Woo woo woo woo-woo-wooo," T-Paul said in his typical conversational holler.

"I hear ya," Bip said.

"Hey, Bip!" I called. "I want to give this wine to my mom for her birthday, so it has to be a surprise. Would you please sign for it?"

"Wellll, I ain't so sure," Bip said, scratching behind his ear. "But okay, if it's for your mama."

He scribbled his name on a very unofficial-looking piece of paper, and I was off! Off with my bottle of sweet, sweet Beaujolais! The only person I could find to share it with was Joy Fink. We weren't exactly buddies.

"Sarah, I've compiled a list of reasons you are not invited to my birthday party," Joy had said a few weeks ago in front of the entire class, handing me a sheaf of sherbet-colored stationery. "Also, I don't like you."

Today she wore a cotton lisle lilac Polo shirt, collar turned up, and matching lilac pom-pom ponytail holders in her hair. She seized the neck of the bottle.

"We have to go drink this," Joy said, scanning the horizon, wild-eyed.

A few blocks away, we found a van with its rear door unlocked. We climbed in. I carved out the stopper with a greasy screwdriver. We lay on the floor of the van and chugged the wine down, cork bits and all.

"You're my best friend in this world life," I slurred to Joy.

We rode the umbrella ride ten times straight. My treat: Mom was running the doll booth at the fair and I stole some of the tickets she had collected. When we got off, Joy clutched my arm like a best friend would.

"Sarah," she said, dizzy and panting. "You're still not invited to my birthday party."

✵ ✵ ✵

The night before we left on our spring break trip to Disney World, my parents had another fight. Dad poured a drink on my mother's shirt as she was breast-feeding Michael. Mom retaliated by going out in the driveway and hurling a potted plant at the windshield of the red Chevy Impala. The pot was an old ceramic Pfaltzgraff jar, the largest of a set of four, meant for flour. It had a blue numeral 4 on it. The plant in it was long dead.

Early the next morning, we loaded into the van to leave for Disney World. Mom was in the front passenger seat, holding a cloth diaper filled with ice on her eye.

"What's wrong with you?" Dad asked, aiming the air conditioning vents up at the ceiling. He always said the car would cool off faster that way.

We pulled out onto the highway.

"I think I have a scratched cornea," Mom said, Michael latched onto her breast.

"You want me to turn around?" Dad asked.

My sisters and I dozed off in the bench seats.

A few hours later, Mom said, "I can't see out of this eye."

"Do what?" Dad said. "I asked if you wanted me to turn around over sixty miles ago. God damn it!"

Dad jerked the wheel hard to the left, executing a u-turn across the median of Interstate 12. The interior of the van became a vacuum that sucked the breath out of my lungs.

Back in our driveway, we filed out of the van. I felt embarrassed, but I didn't know why.

Mom ran inside the house, clutching Michael to her chest, her nipple still in his mouth.

Dad snapped his wad of grape Bubble Yum and said, "All right now, who's going to Disney World with me?"

Hannah eyed the purple gum and slunk off after Mom.

"Well?" Dad said.

After some negotiations and suitcase reorganization, it was settled. Hannah and Michael and Mom were staying home. Becky, Debbie, and I were going with Dad. This was the same way our family was divvied up in my dreams. When the Nazis lined us up in the concentration camp, Mom and Hannah and Michael always got to go free, while the rest of us were sent to the gas chambers.

That night we hauled our luggage into a single room at Howard Johnson's.

"You're sharing a bed with Dad," I told Debbie.

"Yeah!" Becky hissed.

Debbie mashed her lips together, folded her arms, and sat on the edge of the chipped bureau.

Dad called Mom to tell her our room number. He was quiet, listening for a long time.

"I bet you're happy about that," I heard him say.

He hung up.

"This afternoon your Grandma Vivian vomited into her own lungs. She had to be moved from Sunset Plantation to St. Tammany Parish Hospital."

Vomited into her own lungs? First of all, it sounded impossible. Second of all, it sounded like the worst thing in the world.

"Are we going home?" I said timidly, hoping I wasn't giving him any ideas.

"Her condition's stable," Dad said, putting his socks in the bureau drawer. "We might as well finish out our vacation. Already paid for the Disney World tickets."

We all took a deep breath. The hotel room smelled like Grandma Vivian: cigarettes and Final Net hairspray.

"I'm going out for a walk, stretch my legs," Dad said.

Debbie and I lay on separate beds, watching the local news. A dryer on someone's porch had exploded, catching their house on fire and burning it down.

"I didn't know a dryer could explode," Debbie said.

Dozens of neighbors talked into the camera, giving their side of the story. Everyone had an opinion.

Becky came out of the bathroom.

"I got my period," she said, white as a sheet. "I didn't bring any granny pads."

Dad came back with a bag of ice and a bottle of Jack Daniel's.

"Dad," Becky said, "I, I, I . . . "

"Do what now?" he said, unwrapping one of the plastic hotel glasses.

"I need some . . . I got my period."

"WHAT? YA NEED TAMPONS?" Dad said, his volume not jibing with the nonchalant tone he was going for.

"No," she whispered. "Gran—*maxi*-pads."

"Maxi-pads? Awright," he said, tossing the still-empty glass on the bed.

Becky woke up the next morning broken out in chicken pox. She had babysat for some scabby-looking kids a few weeks before. Their mom swore they weren't contagious. There was nothing to do about it now. She languished in the hotel the two days we did Disney, but on the third day, she rose again.

Even a fever of 104 degrees Fahrenheit couldn't keep Becky from the seventeen waterslides at Adventure Island. Everyone stared at the dime-sized scabs showing through her wet t-shirt. When she sat down with me at the snack bar, I scooped up my chili-frito pie and moved to another table, joining a puzzled black family.

"Jeez, didja see that one over there?" I said, rolling my eyes toward Becky, mistaking them for people who would appreciate derisive comments about someone's skin.

The next day, Becky stayed in the hotel room while the rest of us went to Sea World.

* * *

"Who wants to come to the funeral home with me?" Dad called, walking up and down the hall.

Grandma died in the hospital a few days after we got home from Disney World.

"I'll go," I said, wondering why I didn't feel more: more angry at him, more sad about Grandma, nothing.

"Would you like a soft drink?" the funeral director asked me as we stood under the gigantic chandelier in the foyer.

He was tall and ghostly, with slicked hair and steepled fingers: just how you'd expect a mortician to look.

Dad and I were ushered into a casket showroom. We moseyed around like we were wasting time at a car dealership. The most expensive casket was splayed open right up front on a pedestal. There was a sign detailing its appointments: rich rubbed mahogany and cream silk satin interiors.

I stopped in front of an ombré lavender coffin. The handles were white ceramic enamel painted with pink roses.

"Grandma would like that one, wouldn't she?" Dad said from right behind me, making me jump.

Would she? I didn't know what Grandma would like. Where was her body, anyway? Somewhere in this building? I felt blood rushing away from my head and heard a ringing in my ears.

"Do you feel strange?" Dad asked me.

I willed myself not to faint.

"I have to go to the bathroom," I said brightly.

I ran through the big double doors back out under the chandelier, which pelted me with light. The funeral guy was standing there.

"Care for a soft drink?" he said.

"Where's your restroom?" I said, holding my thighs and swallowing down vomit.

"To the left," he pointed with his long, long index finger.

In the bathroom, I splashed water on my face just as they did in made-for-TV movies. I wanted to press my forehead against the grey marble wall, but I didn't. I felt there was a good chance I was being watched. Someone tapped at the door.

"Uhm, yes?" I said.

The funeral director poked his head in.

"Are you sure I can't get you a soft drink?"

That night Dad drove me and Becky up to a football game at Saint Jude's, the boys' school in Covington. Becky hung out with her friends Elizabeth and Elaine, the other two smart girls in her class. I turned up the collar of my turquoise Esprit polo shirt and flirted with Danny Cutrer, a rich kid two years younger than I. Dad sat up in the bleachers, pulling on a fifth of Jack Daniel's.

✵ ✵ ✵

The funeral was in Kansas City. Dad flew up with Grandma in her lavender casket. The rest of us stayed in Louisiana.

When he got home, Dad didn't unpack his suitcase. He made a ramp out of plywood up to the bed of his truck. Our next-door neighbor came over and helped him push his giant roll-top desk up the ramp. They tied it down with yellow nylon rope and bungee cords.

"I'm moving out," he said. "I'll see you next weekend."

Mom was inside the house with Michael. I stood with my sisters beside the truck. Dad got in behind the wheel and put the truck in gear.

I wondered if we had to stay here until he was all the way down the driveway.

We had a very long driveway.

Once his truck was out of sight, we jumped up and down in slow motion, weightless.

Free at last, free at last.

Top o' the Food Chain

"After many failed business ventures, my father died, leaving my mother and sisters and I penniless," I said in a clipped British accent. "*I must have something to engross my thoughts, some object in life which will fill this vacuum, and prevent this sad wearing away of the heart!*"

I flung myself on the floor before my eighth-grade English class at St. Aldric's, wearing a muslin bonnet, a hoop skirt covered with burlap, and an eye patch. The assignment had been to present a report on a famous person. Mom suggested that I dress up like Elizabeth Blackwell, the first woman doctor.

I wanted satin and velvet, but Mom insisted upon fabrics appropriate to the 1800s. She made the hoop skirt beneath out of an old sheet and a hula hoop. Whenever I moved, the beads inside the hoop made a *shuk-shuk* noise.

"I was born in England but lived much of my life in the United States," I said from the floor, reading off index cards. "I later studied with midwives in Paris, France. Ah, Paree! It was there I contracted a horrible infection that left me blind in one eye."

Daintily removing my glasses, I pulled up the eye patch and bulged my eye out extra big, rolling it 'round like I'd seen Marty Feldman do in *Silent Movie* on the local UHF channel.

Mickey Foote raised his hand.

"Yes, young man?" I said, pointing at him.

"If your eye was so bad," he said. "Why didn't they dig it out and throw it away?"

"Uh, well, believe you me, buddy, they tried!" I ad-libbed. "But I wouldn't let them. '*Nay!*' I said. '*You will not take my eye, you rapscallions!*' I said."

I got up from the floor—*shuk-shuk*—and referred to the cards for another direct quote.

"It is not easy to be a pioneer—but oh, it is fascinating! I would not trade one moment, even the worst moment, for all the riches in the world!"

"Are you like, dead?" Andy Bourgeois said without raising his hand.

The only boys left at school were the ones too poor to go to eighth grade at St. Jude's. They were troublemakers or dumb or both.

Tim Piles lifted a clear plastic Bic pen casing and shot a spitball. It landed on the lens of my glasses, sticking for a moment, then slowly sliding off.

"Our school education ignores, in a thousand ways, the rules of healthy development," I said primly, glancing at my teacher, Mrs. Strain. She was smiling vacantly out the window, probably dreaming of her next cigarette. "In 1907, I fell down some stairs, ending my career as a gynecologist."

The dismissal bell rang.

"Great job, Sarah," Mrs. Strain said. "Could I see you for a minute?"

"WhoooooooOOOOOOOO," hooted the class, like the studio audience of *Good Times*.

Nervous, I shoved the bonnet, eye patch, and cue cards into my desk. I walked toward Mrs. Strain—*shuk-shuk, shuk-shuk*.

"The boys quit the play. They think they're too cool for it," she said, lasering me with her perpetually cocked eyebrow. "Do you want to play Hubs, the leader of the motorcycle gang?"

Mrs. Strain was in charge of the annual eighth-grade play. This year it was set in the 1950s, a *Bye-Bye Birdie* rip-off called *Teeny Boppers*. The dialog was a hybrid of *Grease, Gidget,* and *Palm Springs Weekend,* lots of "Corndoggie" this and "Chickie" that. I'd originally auditioned for Janet, the female lead. I couldn't figure out why Mrs. Strain cast me as Cheerleader number 2.

"Of course!" I said, finally getting the meaty role I deserved. "I think I'd make a great Hubs."

"I think so, too," she said, cuffing me on the shoulder and double-clicking out the side of her mouth.

I paid close attention as Mrs. Strain swaggered out the door, rolling on her haunches. Hers was just the kind of macho/sexy vibe I could incorporate into my characterization of Hubs, leader of Da Wheels.

The night of the play, Mrs. Strain loaned me her own black leather cap and a pack of Salems to roll up in the sleeve of my white t-shirt. I stuck one behind my ear and let another dangle out of the side of my mouth. Its minty flavor was pleasurable. I wished my parents smoked menthols.

"Hey Chickie, what's your name?" I said, terrorizing the poodle-skirt-wearing goody-goodies at the malt shop.

I stole the show. I *saved* her show, and how did Mrs. Strain repay me? By giving me a C on my Elizabeth Blackwell report.

"That costume alone was worth a B+ at least," Mom said.

* * *

"Did your dad hit your mom before or after she threw the plant at his windshield?" asked Mr. Frey, my dad's lawyer.

I crumbled inside. And when I realized I was crumbling, I cried and I couldn't stop. It was the crumbling that made me cry, not the topic of conversation.

"Okay, let me ask you something else," Mr. Frey said, fake-softly. "Who cursed more, your mom or your dad?"

"I, I, I . . ." That was me, blubbering.

"So they both cursed, huh?"

"Let's take a short recess," the judge said, looking at me with pity in his eyes. He handed me a box of Kleenex.

Don't feel sorry for me, butthole!

Sympathy always made me cry harder. I stepped down off the stand. Becky and Mom and I went into a little room with her lawyer, Mr. Taylor.

"Sarah, sugar, it'd be real helpful if you could get a hold of yourself," he said.

"Becky already testified. Isn't that enough?" Mom said.

Becky watched me, her mouth pressed into a disapproving line.

"I thought you said you were gonna tear into Dad up there," she muttered. "Well, *that* was a great plan."

Fine, I would tear into him on the *Tonight Show*, whenever I ended up on it. Though I'd rather do it on *Late Night with David Letterman*. I liked him better than Johnny Carson. Things were not always what they seemed on the David Letterman show.

"I'll do it on David Letterman," I promised Becky quietly.

"Ooooh," Becky said. "That's perfect."

✲ ✲ ✲

"Jesus Christ!" Mom said, rushing into the house with blood spattered all over her bare legs.

"What? WHAT?" I said, peeking out from behind my fingers.

She'd been out mowing the lawn, something she did since Dad moved out. She'd go out in a blouse and wrap skirt and push the mower, her gaze steely and her body at an almost 45 degree angle to the ground.

"God damn," Mom gasped. "I ran over a turtle."

That explained the blood on her legs. Our lawn mower didn't have a bag attached to it, so whatever was run over—grass, twigs, wildlife—shot out from beneath it, sliced and diced.

I ran out to the front yard. Under the mimosa tree there were pieces of shell and bloody guts and scaly legs with tiny toenails strewn about. Gag-a-maggot. I went back inside. Mom was still on the sofa, breathing hard.

"Did you smell it? I could smell it. It made me feel like an animal, that smell," she panted. "I felt like I wanted to eat it."

I went out and raked up the turtle pieces as best I could. Before Mom decided to eat it raw, or worse, cook it up for dinner. I smelled what she was talking about. It killed my appetite entirely. It smelled like a used granny pad. I was hoping there would be enough salvageable shell to make an ashtray or at least a medallion, but no luck.

Later that night we were watching *Monty Python's Flying Circus* on PBS. There was a long cartoon of women with trombones and tubas and other brass instruments stuck onto their breasts. A giant foot came down from the clouds and squashed them.

"I still feel like an animal," Mom sighed, tossing a handful of popcorn into her mouth.

✵ ✵ ✵

"Boi—yoi—YOI—YOING!"

The cartoon penis shot up incrementally into an erection. Everyone in the classroom cracked up.

"Can it," Mrs. Strain said gruffly.

The new thing at school was sex education or, to be more specific, sex coeducation. They were showing us, girls and boys together, a film that purported to explain it all. Afterward, the town pediatrician Dr. Camp offered to answer any questions we might have. One person, Jacques Lala, raised his hand.

"Do guys ever run out of spoim?" he said, his pronunciation the same as an eighty-year-old Cajun's.

"Ho ho ho," Dr. Keller chuckled. "No, we men have plenty of sperm, up until the time we're ninety-nine. It's the gals' eggs that peter out."

Everyone laughed. That was the end of the discussion. No one wanted to appear ignorant, or worse: curious.

"Jacques wants to go with you," Vito Costanza said later on the playground. "Do you want to go with him?"

To "go with" someone meant to go out with him, to date, go steady. I pondered the offer.

Jacques wore thick black-framed eyeglasses decades before they were the mark of ironic sensitivity. His fastened around the back of his head with a snapping strap. His hair was shorn into a glossy, mouse-colored crew-cut. Jacques was a runt, but I was on the short side, too. He pronounced three "tree," but that was something we could rectify. Besides, no one else had ever been interested in going with me before.

Through a network of middlemen and -women, Jacques and I agreed to go with each other. Our first date would be a big outdoor party at Travis Kehoe's house out in Folsom.

"I don't like these boy-girl parties," Mom said, driving me over there.

"Mom, everybody goes, it's no big deal," I said.

"Celeste Hazard's mother's not allowing her to go," Mom said, turning into the Kehoes' driveway. "I think she's got the right idea."

"Celeste wasn't even invited," I said.

Celeste Hazard had once been my best friend—BRIEFLY. I was over at her house when John Lennon died. Her whole family gathered to watch the news in the family room. Celeste's two older brothers lay on the floor, crying.

"Why?" they sobbed. "WHY?"

I was trying to muster up a tear or two myself when their dog, Smokey, a lab-shepherd mix, put his paws on my shoulders and started to hump me from behind. I could feel his grody pink dog penis poking into the small of my back, like an insistent roll of Life Savers.

"Smokey likes you!" giggled Celeste's father.

Her brothers laid aside their grief, rolling over on their sides to enjoy the spectacle of dog-on-girl rape.

"I miss Celeste," Mom said, pulling to a stop in front of a giant house. "There's no cute boys in your class anyway. Nobody as handsome as John Boy."

"Gross!" I said, kicking open the car door. Leave it to my mother to still be mooning over the guy with a big mole on his face from that TV show *The Waltons*.

The Kehoes' house looked like an antebellum mansion. Travis was only a seventh grader, but his prematurely hairy legs allowed him to go with an eighth-grader, Malison Drucker. She greeted me on the sprawling, wraparound porch with a baby bottle full of something light brown.

"Wanna sip?" she said, twisting her rat-tail around her finger. "It's sherry."

I grabbed the baby bottle and took a slug. More like a suckle. It tasted like a saltier version of Mom's Steak Diane.

"Harvey's Bristol Cream, eh?" I said, showing off the booze knowledge I'd gleaned from commercials.

"I dunno. I just snuck it out of a bottle in our pantry," Malison said.

I walked into the living room. Some kids were couple-dancing at arm's length to Journey's "Lights." Joy Fink walked by, drinking out of a Vidal Sassoon conditioner bottle.

"Jacques Lala's looking for you," she said with an evil smile, flashing the pointy bicuspids up high in her gums. "I'd offer you some Amaretto di Saronno, but I'm saving it for my real friends."

I felt sick to my stomach.

I ducked into the bathroom to put on some frosty blue eyeliner. My mother would never have driven me here if I'd been wearing makeup. I leaned over real close to the mirror, pulled my lower lid out, and sketched it in.

I went back out into the yard and stood with Michelle Banks and Ellen LaCour next to a giant twisted oak.

"Did you talk to Jacques?" they asked.

"Not yet," I said, cracking my knuckles. "I have to break up with him tonight."

"Yes, you totally should!" they said.

The beginning of a Foreigner song drifted out of the house into the yard.

You're as cold as ice
You're willing to sacrifice our love

Vito Costanza ran up, sloshing a Boy Scout canteen full of gin and Coke.

"Jacques told me to tell you he doesn't want to go with you anymore."

"Tell him I already broke up with him," I said. "But we can still be friends."

✵ ✵ ✵

In the divorce settlement, Mom got no alimony. I guess the judge had determined my parents cursed in equal amounts.

In addition to $150 a month child support for each of us, she got half the money from Dad's profit-sharing and pension plan in a one-time lump sum. When the check came, we cashed it and lit out for Panama City Beach, Florida. The waves were hard and grey. They left rippled bruises on Mom's thighs.

"I bruise easy," she said. "Low iron."

My sisters and I all ended up stung by jellyfish. We lay on the beds in our hotel room, baking soda paste caked on our legs. The ceiling started leaking rusty water onto us.

"Mom!" we carped. "Ask the manager for a refund!"

"Calm down," she answered. "When we get home, I'm going to buy you each a pair of red shoes."

Becky got a pair of red flats with scalloped edges, mine had a crepe sole with crisscross straps, and Hannah and Debbie got matching patent leather Mary Janes. Michael got little red baby Keds.

The divorce settlement also decreed that our house in Piney River Country Club go on the market. Mom and Dad would split the proceeds from the sale.

The house sold pretty fast, bought by a couple with three children. We spent the months leading up to our move-out date stacking, rather than packing. We couldn't afford to buy boxes just yet, but there was no use putting clothes back into drawers or books back on shelves, when we would just have to take them out again to pack.

"What are you doing?!" Becky asked, walking into my parents' bedroom—former bedroom—one afternoon a couple of weeks before we were supposed to vacate.

"Mom said we could," Hannah said, holding up the purple marker in her hand in self-defense.

On the bedroom wall, Hannah and her friend Claire were drawing a four-foot-tall moose wearing a nun's habit and sitting on a toilet. Another, smaller moose was holding a toilet brush aloft like a scepter. On the opposite wall, Debbie was punching the buttons on a cash register I had expertly rendered.

"This is sick," Becky said. "What will those people think when they move into this house?"

"Aw, come on, it's fun," I said, shading in a giant swastika.

I had a little Nazi Fever hangover. My eighth-grade Social Studies Fair project was a thorough explication of Zyklon B, the crystallized cyanide gas used in the gas chambers at Auschwitz. I was obsessed with the minutiae of Nazi atrocities. If I was up on all the statistics, I could make sure it never happened again.

I held a marker out toward her, a nice fat black permanent one.

"I WILL NOT!" Becky said, hysterical. "If this gets out, everyone's going to think we're crazy!"

"Come on," I said. "It's not half as bad as the time Hannah drove her Big Wheel down the street with no underwear on."

"Shut up!" Hannah said.

Claire laughed.

"That never happened," Hannah said to her.

"Why are you so obsessed with swastikas?" Becky asked me. "That's all you ever read about: Hitler, Nazis, death camps, gas chambers, ovens, lampshades made of Jewish people's skin!"

"I read other stuff," I said. "I just reread *Helter Skelter*, remember?"

I looked for a blank space on the wall where I could scrawl "PIG."

"Right, how could I forget?" Becky said, launching into an extremely lame imitation of me: "'Patricia Krenwinkle stabbed Voytek Frykowski fifty-seven times! They cut open Sharon Tate's stomach and took out the baby! Squeaky Fromme escaped once; what if she does it again? Wahhhhhh, I'm scared!'"

"That's not what scared me," I said. "It was that those people didn't even KNOW those people and they just went in and killed them!"

"I don't care. If you're gonna read books like that and get all scared, it's your problem," Becky said. "Don't come crawling into my room, begging to sleep with me anymore."

"Fine," I said. "Find someone else to tickle your back every night!"

Groalp!

We all froze. The sound came from out on the street, accompanied by the screech of tires. We recognized that *groalp*.

Our dog Beau had been hit by a car.

"Can I give you some money?" the driver asked my mom.

"No, don't worry about it," Mom said. "He'll be fine."

Beau had internal bleeding and a collapsed lung. They had to shave a patch on his back, stick a needle down into his lung, and re-inflate it. He came home from the vet, red-eyed and lethargic.

"The vet says Beau's depressed," Mom said.

She paid the vet bill with our child support. We pooled our babysitting money and piggy banks and ate fried eggs for the rest of the month. Now we were all depressed.

✵ ✵ ✵

We saw Dad every other weekend. Either he'd pick us up and take us back to his apartment in Baton Rouge, or we'd go to the park down in Mandeville by Lake Pontchartrain. We were never much of a park-going family before the divorce, so we weren't quite sure how to behave there. Lakefront Park consisted of a couple of concrete tubes scattered around a sandpit and a rusty chin-up bar. We'd sit around for five to six hours, on or in the tubes, watching people gliding in and out of the Yacht Club for brunch.

One afternoon around 4 o'clock, as soon as supper could be justified, we headed over to Pizza Inn. We loaded up the jukebox with our favorites and sat down to order one large pepperoni and one medium pork topping with black olives.

"So, high school," Dad said, looking at me. "A big step."

"Yeah . . ." I said, trailing off and tapping my fingers to "Stumblin' In" by Suzy Quatro, Hannah's pick.

Grade school was drawing to an end for me. For two years already, Becky had been up at the all-girls Catholic high school, Our Lady of Prompt Succor. She was making straight A's and the kind of friends that go with them: tall, short, Chinese, and red-headed.

"I'm glad your mother agreed to send you girls to public school," Dad continued. "We can't afford parochial."

"Yeah . . ." Becky said.

Debbie was dramatically lip-synching the opening monologue of "I Will Survive":

At first I was afraid, I was petrified.

"I suppose you'll be needing some new clothes, seeing as you won't be wearing uniforms anymore," Dad said. "I think we can take care of that."

Unbeknownst to him, Becky and I weren't going to public high school. Mom would rather die than send us there. Last week, she preregistered us both at Our Lady of Prompt Succor. We'd be wearing uniforms.

"Keep this from your father for as long as possible," Mom had said. "Public high school's full of sluts—there's no way you're going there. Don't worry, I'll get the money together somehow."

I wouldn't mind some new clothes. I still adored Lacoste, from afar. Maybe I could get that pair of pinstriped Guess? Jeans with the zippers at the ankles, and those navy polka-dot linen Perry Ellis flats from Gotcha Covered, the boutique next to Walgreens.

"Chuck E.'s in Lo-ove," I sang, using one hand to straighten my imaginary beret and the other to hold a drinking straw cigarette. I kept my eyelids at half-mast like Rickie Lee Jones did when she sang it on *Saturday Night Live.*

"We'll go shopping, get y'all some school clothes," Dad said.

"Great," Becky said, picking up the shaker of crushed red pepper and holding it like a microphone. While she waited for the opening chords of her song, "Sad Eyes," to segue into the lyrics, she absentmindedly lifted the shaker up to her nose and sniffed it.

"AAAAAAYYYYYYYY," she cried, tears streaming down her cheeks.

Becky spent the rest of the meal in the bathroom, trying to dig and blow the red pepper flakes out of her nose.

The check came right when Dad's song started, Neil Diamond's "Forever in Blue Jeans":

Money talks,
But it don't sing and dance and it don't walk.

"Here ya go, hon," Dad winked, handing the waitress thirty dollars on a $28.75 tab. "Keep the change."

Dad had always flirted with waitresses, giving them lingering grins. Now that he was technically single, it had a sinister air. The moustache he'd grown made him seem like more of stranger than ever.

As long as I can have you here with me
I'd much rather be
Forever in blue jeans, babe.

When Dad lived with us, one of my jobs was to iron his jeans. I dropped a napkin and ducked down to get it, sneaking a peek. They still looked ironed. Who was ironing his jeans now?

The waitress walked away from our table, counting the money. She looked over her shoulder, winking back at him. Was *she* ironing his jeans?

That night Dad dropped off Hannah and Debbie at home and took me and Becky back to Baton Rouge to spend the night.

"Dad got cable!" I whispered to Becky, changing the channels on the tiny TV that used to belong to Grandma Vivian.

Dad came out of the bathroom and took the remote out of my hand.

"Let's see what's on HBO," he said.

Private Benjamin. I was so excited to be allowed to watch an R-rated movie, I almost forgot what made a movie rated R.

Goldie Hawn got married to some annoying guy. I knew what was coming next: Adult Situations. Strong Language. Nudity.

"I've got to go to the bathroom," I mumbled, right as the groom mounted Goldie in their honeymoon bed.

I dove over Dad's legs, leaving Becky alone with him to witness the sextacular.

I stayed in the bathroom a long time, flushing twice.

When I came back, Goldie was in the army. Her annoying husband was nowhere in sight.

"Thanks a lot," Becky whispered, gouging her fingers hard into my thigh when I sat next to her on the sofa.

"I had to pee," I said innocently.

We watched the movie a bit longer, until my curiosity got the best of me.

"What happened to her husband?" I asked.

Silence.

"What happened to her husband?" I asked twice more.

"He had a heart attack and died," Dad finally said, "when they were MAKING LOVE."

✵ ✵ ✵

With Dad gone from our house, my social life improved. I could attract the sort of people who had previously been reluctant to sleep over: ladies of leisure who preferred to stay up late and sleep in, rather than get up at dawn and rake the yard with a working smile. Plus, if I had an overnight guest, I could get out of spending the weekend at Dad's apartment in Baton Rouge. His crushed gold velour sleeper sofa and non-stick skillet had quickly lost their novelty value, and I couldn't risk another HBO debacle.

I cozied up to Donna Grunditz. She had just moved to St. Tammany Parish from Dallas. She wore green eyeliner on the inside rims of her hazel eyes and had the biggest boobs I'd ever seen. Those boobs could elevate the status of anyone. Last year, Buck Gurley was just another abnormally tall kid whose neck and head had the same circumference. Since he started going with Donna, he was a first-stringer on Mandeville Junior High's flag football team. Donna wasn't stuck up at all about her boobs. She was quite willing to use them to lure guys for her friends, too.

Asthma drugs had stalled the onset of puberty for me, a sort of pharmaceutical ballet. Lithe and hollow-boned like a bird, I still hadn't gotten my period. My concave chest could pass for a set of chip and dip bowls.

"God, my back hurts!" Donna said, lifting her full C-cups through her uniform blouse. Clasping her hands together in a prayer position, she rubbed the space between them. "You wanna sleep over Friday night? Buck's sleeping over at Flip's and we can, you know . . . sneak out?"

"Sneak out of what?" I asked, wondering if this was a stupid Texan misunderstanding of the song "Freak Out."

"Sneak out and meet the guys, dummy!" Donna laughed.

Friday afternoon I went home from school with Donna. Her mom sat in their sunny dining room, painting her nails.

"Nice to meet you!" she said. "Donna, it's time for our snack. You hungry, Sarah?"

"Sure," I said. They seemed like the type of people who always had scads of unopened, never-fought-over, ever-crispy snacks. The type of people who said, "Time for our snack," rather than "Gimme that!"

Donna came out of the kitchen carrying a plate of three hot dogs in buns and a pile of potato chips, and two small bowls.

"Here ya go," she said, setting the plate down in front of me.

She and her mom each took one of the small bowls.

"We're on Weight Watchers," Mrs. Grunditz explained, digging into a tiny scoop of chicken salad with a tiny fork, her wet nails carefully fanned out. "But you go right ahead, don't feel bad! You want mayonnaise?"

Mayonnaise on a hot dog? Mayonnaise on anything? I hated mayonnaise. I hated everything about it: its slick gurgle and curdy texture, its tendency to turn clear when smeared on a plate and left out in the sun. Once my sister Hannah chased me around with an open mayonnaise jar and I dove out the kitchen window to escape. Lucky for me, I landed in one of our wild, cushiony, untrimmed hedges.

"No, thank you," was all I said.

"I can't live without my mayonnaise!" Mrs. Grunditz said, swirling some onto a rib of celery. "Donna, get Sarah some Thin Mints from the freezer. The Girl Scouts came by yesterday and I can never say no. But darn, I need to get those out of this house before they drive me to distraction. Sarah, you just sit down here and eat as many as you can."

I sat at the table snapping frozen cookies in my mouth while Donna's mom pulled her hair through little holes in a plastic cap, daubing bleach on the strands with a teeny paintbrush.

"A girl needs her highlights," Mrs. Grunditz said, holding one of those new, super-skinny cigarettes in her plastic glove. Taking a drag, she continued, "I'd LOVE to get my hands on your head of hair, Sarah. Bring out some auburn streaks."

While the bleach did its work, Donna and I went up to her room and listened to a Blue Oyster Cult *Greatest Hits* cassette.

"My mom loves you because you're smart. She's hoping you'll raise my grades," Donna said, slipping on a pair of control-top suntan pantyhose and polishing up her boobs. "I can't WAIT for tonight."

"Do you think it's safe out there?" I said.

"Don't Fear the Reaper" was making me feel dreamy and wistful and dead, dead like Faith Hathaway. Faith Hathaway, a local girl, had been kidnapped a few years ago. They found her naked body, raped and stabbed, out in the woods.

"Her hips were pulled clean out of the socket," Mom would always say right as I gamboled out the door, looking for fun and adventure. "Just because they got Willie up in Angola doesn't mean you shouldn't keep an eye out."

She was referring to Robert Lee Willie, the guy sentenced to death row for Faith's murder.

It was as if she had embedded a computer chip in me. Wherever I went, I first made a cursory scan of all male faces for the intentions to rape and murder. Then I could half-relax and semi-enjoy myself.

"Sarah, we'll be with the GUYS," Donna said. "They'll totally proTECT us."

"Let's listen to the B-52s instead," I said, looking at the band photo on the album cover. "Don't you think the lead singer would be pretty cute if he shaved off that moustache?"

That night at 11 o'clock, we climbed down the aluminum gingerbread column that supported the balcony off Donna's bedroom. Down the road in a wooded lot, Flip and Buck and a third guy, Billy Sticker, were waiting for us. Buck and Donna immediately latched onto each other, open-mouthed, and disappeared into the bushes.

"Let's go to Time Saver and get some beer," Flip said.

Time Saver was a convenience store a few blocks over. Just the mention of it gave me a little frisson of nausea. Their commercials featured pudgy sausage fingers wearing giant rings, helping themselves to fountain drinks and premade sandwiches while a voice-over said, "Hey Anna Mae, I hear Time Saver makes their own my-nez fresh every mawnin'."

"My-nez" was Yat for mayonnaise. "Yat" was the derogatory term for some New Orleanians, derived from their idiom for asking how someone was doing: "Where y'at?" Everything at Time Saver was slathered with my-nez, maybe even the beer.

I looked over at Billy Sticker. He went to public school and was a year younger than me, but had been held back two years. A Little Hercules uni-brow spread low on his forehead like a thorny black caterpillar. Nice lips. I reckoned I could face the my-nez for him.

"Time Saver saves time / Time Time Saver Come ON," I chanted, in a low, threatening voice.

That was their catchy radio jingle.

When we got to Time Saver, we waited out front for a promising bum to come by, someone we could ask to buy booze.

"Okay, first base and second base, those I can understand," I said, making conversation. "But third base? What good is that to anyone?"

"It's pretty good for the girl if you're doing it right!" Flip said, laughing and butting chests with Billy.

"Look at that dude—he'll buy it," I said, watching a guy weave his way past the gas pumps, stopping to count something imaginary in the distance with his index finger.

"You ask him," Billy said, his eyebrow suddenly shrinking into a tweezed-looking, feminine arch.

Flip smiled winningly at me. I felt a hard core swelling inside me, manly. I'd get them boys their booze.

After some small talk with the bum, money and beer changed hands. We ran back to the woods with our six-pack of Moosehead. We were too young to understand quantity was better than quality. Or which brands didn't have twist-off caps.

"Shit," Flip said, cutting his tongue while trying to open a bottle on his braces. "Why didn't you tell that dude to buy cans?"

For a split second, I panicked, feeling the bonhomie of our time together at Time Saver wearing off. Buck and Donna rustled and moaned in the bushes.

"I'll go back to Donna's and get a bottle opener," I said, shooting a look at Billy that said, "Come with me and you might just get to feel my boobs. *Over* my shirt."

"Okay," he and Flip said.

I guess it was too dark for Billy to catch my drift. I ended up going alone. Should I shimmy back up the gingerbread column to Donna's room? I tried the front door. It wasn't locked. That was pretty dumb. Hadn't Donna's parents ever heard of Faith Hathaway? I would have to have a little talk with them in the morning.

I stepped into the dark foyer and began to close the door ever so gently, leaning against it to ward off creaks.

"Where's Donna?"

It was Donna's mom, standing behind the door.

"GAHHHHH–hello," I said, coasting from a high-pitched murder-rape victim's scream to a casual, retarded robot tone of voice. "She. Out. Side."

"Is she with Buck?" Mrs. Grunditz said, swishing toward me in her matching polyester charmeuse gown and robe.

"Well, I not sure, maybe," I stammered. "With Buck and some other. People."

"Oh. OK," she said.

"Well!" I said, clapping my hands once and brushing past her into the kitchen.

I turned on the overhead fluorescent light and began rooting through the drawers looking for a bottle opener. It seemed like the thing to do. The boys had asked me to go get a bottle opener. They were waiting. They needed their liquor. I needed some liquor. I still wasn't buying into the hype about third base, but I wouldn't sneeze at getting to first.

I found a bottle opener and went back to the front door. Mrs. Grunditz was still standing there, staring at me.

"Tell Donna she better get in here right now."

I walked out into the yard and down the road unhurriedly, running the rest of the way as soon as I was out of sight. I tossed Billy and Flip the bottle opener and rushed into the bushes, where Donna and Buck were dry-humping under a blanket of decaying leaves.

When we got back to Donna's house, Mrs. Grunditz didn't yell or anything.

"We'll discuss it later," she said calmly, pointing at us with all ten of her long, painted fingernails.

Up in Donna's room, I wondered who would be discussing what with whom.

"Is she gonna tell my mom?" I agonized, more worried about Mom telling my dad.

"Probably not, she's cool," Donna said, looking in the mirror at her fresh hickies. "My mom's my best friend."

Early the next morning, I got up to go to the bathroom and stepped on something sharp in the brown shag hallway carpet. It felt like a shot going into the ball of my foot.

"I think it's fine. You probably just got pricked by a carpet tack," Mrs. Grunditz said, looking at the tiny dot on my sole. "By the way, I've decided not to tell your mother about last night. Now listen: you hear that? Girl, you got a stack of pancakes waiting for you downstairs, crying out for syrup!"

She and Donna ate chicken salad on dry toast.

By the following weekend, I was walking with a pronounced limp. The entire bottom of my foot was swollen and tender to the touch. Saturday morning my dad came to pick us up and take us back to Baton Rouge.

"What are those fried eggs doing out in the driveway?" Dad said.

"Somebody egged our house last night," I answered.

It was almost a hundred degrees outside. The eggs had cooked on the pavement.

I was lying on the sofa with my foot propped up, a cold rag on my head, sipping 7UP.

"I can't go to Baton Rouge," I said. "I stepped on something sharp at Donna's last weekend."

"Lemme see that," he said.

Dad lifted my foot and squinted at it.

"There's something still stuck in there," I said.

Dad took his broad thumb, a crescent of grease under the nail, plunged it down onto the ball of my foot—the epicenter from which the pain and swelling had spread, ground zero—and rubbed vigorously.

"GAHHHHHH," I said.

My body tried to double over. Because of Dad's grip on my ankle, I was only able to flip onto my stomach and flail like a fresh-caught fish.

"Nothing in there," Dad said, dropping my foot back down on the sofa.

Finally, when I had to resort to crawling around the house, Mom took me in for an x-ray. There was a needle deep in my foot, shoved right up between my toe bones. Dr. Camp would use the fluoroscope over at St. Tammany Hospital to see it while he dug it out.

"You remind me of my brother Joe," Mom said, driving to the hospital. "He was such a hypochondriac, the one time he really got something, my parents didn't believe him. He was lying in bed with a 105-degree fever, talking nonsense. Come to find out, his appendix had burst right inside him."

Dr. Camp shot my foot up with so much Novocain, it was numb for two days. We couldn't afford to rent crutches, so I lay on the divan, watching *Days of Our Lives* and *Ryan's Hope* and *The Edge of Night*.

"I need more chipped ice!" I called from my sickbed. "And some fresh gauze!"

On the third day, Debbie came home from school with Elizabeth Rhodes. Immediately, I could tell she was a vast improvement over Loma Santos, Debbie's usual companion. Last time Loma came over, she was wearing an old Ziggy t-shirt of Hannah's that Mom had donated to Goodwill. I recognized it from the stain on the front, a yellow splotch of urine from our pet rabbit, smack in the middle of Ziggy's thought bubble.

Darn . . . I'm good! it said.

I used to hang out with Loma's big sister Ada, but I hadn't been over to their house since the time I walked in on their father in the tiny, infinity-mirrored bathroom under the stairs.

"*No no no*!" thousands of him yelled, sitting on toilets, waving their arms, and leaning over to conceal their crotches.

Elizabeth Rhodes had a shiny brown pageboy and twinkling eyes that reminded me of her older brother Hunter, one year my senior and prettier than most girls.

"Elizabeth, I'd love to get you some apple juice, but my foot here . . ." I shrugged. "Could you help me up? I could use some fresh air. It's rather stuffy in here."

It *was* stuffy in there. The central air conditioning had broken a month ago.

"The house is sold. Let them fix it when they move in," Mom said. "I can't waste money on it because I have to pay your Catholic school tuition. Anyway, I for one am glad it's broken. That air conditioner's full of mold; I always feel sick when it's on."

Debbie and Elizabeth helped me up. Gripping their little shoulders like human crutches, I walked out onto the patio, feeling the sun on my face.

"Ahhhhh," I breathed deeply, my eyelids fluttering like the woman who just woke up from a coma on *The Edge of Night*. "This is just what I needed. Thank you."

Debbie looked at me, apprehensive. I could tell she was expecting me to start grating Elizabeth like cheddar cheese.

"So tell me, Elizabeth, how's Hunter?" I said, pulling my Lanz nightgown out at the chest to give the illusion of breasts. "Did he get his learner's permit yet?"

"Look," Elizabeth said, pointing out to the middle of our driveway.

"Eww, a snake!" Debbie said.

I jumped up, momentarily forgetting to be a cripple.

"Salt and pepper king snake. Harmless," I said. "Could be a mock cottonmouth, though."

"I hate snakes," Elizabeth said.

She and I locked eyes.

"Get some pine cones," I said, seeing her and raising her.

When the first pine cone hit, the snake recoiled, slithering toward the monkey grass.

"Cut him off," I said.

When we ran out of pine cones, we used fallen branches, gravel, seed pods, pieces of broken brick from the edging of the plant beds, and my little brother's toy trucks. We lobbed, we spiked, we side-armed. Hannah came outside and silently joined in. Gradually, bits of the snake's flesh were nicked away, blood oozing out from between its scales.

We ran out of ammo and stood around breathing hard.

"It's getting away," Debbie said softly.

"It won't," Elizabeth said, her cheeks flushed. "Let's kill it."

I got the hoe out of the toolshed and handed it to her. The snake had made it over to the side of the house and was trying to burrow under the foundation. Elizabeth gaffed it with the hoe and dragged it into the flower bed. She hacked at its neck with the hoe until its head was nearly, but not quite, severed.

A shiny, late-model station wagon glided up our driveway.

"My mom's here," Elizabeth said, dropping the hoe.

As Elizabeth ran toward her mother's car, I felt as though she were pulling the end of a thread, unraveling me like a sweater. Somehow, I knew we would never see her again.

Movin' on Up

"I wanna be Poor Barbie!" Hannah whined, scratching the scabby volcano protruding from her knee.

"No, it's my turn," I said, setting up the traditional Poor Barbie household on a wooden breadboard from the kitchen: a cast-iron stove and a washrag sleeping bag next to a plastic G.I. Joe campfire.

I was fourteen, but I couldn't resist a game of Poor Barbie. Poor Barbie started when I bought an old doll from a garage sale, the kind from the fifties with real eyelashes and milky white skin. When I got her home and took off her clothes so she could have sex with Ken, I discovered that her smooth breasts and crotch had been scratched and punctured with a straight pin. One of those straight pins with a colored ball on the end, which remained hammered into her butt crack. I called her Susan.

Susan's checkered past made her the perfect mother for our time-ravaged Growing Up Skipper. Unlike the standard, prepubescent Skipper, Growing Up Skipper had a layered haircut and breasts that grew when you rotated her arm. I called her Amy.

I set up Susan and Amy in a one-room house. They lived on ketchup sandwiches and drank warm Coke because they couldn't afford coffee. I shaved both their heads and severed Amy's fingers and toes with a paper hole-puncher.

"My mother burned my toes off with a Bunsen burner in a fit of rage triggered by malnutrition," Amy would say matter-of-factly to the other Barbies, who dropped by for a warm Coke and pork cracklins between fevered assignations with the lone Ken doll in town.

"No fair!" Debbie said. "I was thuppothed to be Poor Barbie thith time!"

"You're supposed to be packing all that up, not playing with it," Mom said on her way through the room, flinging loose clothes and books into a garbage bag. "We're moving out in five days."

Bit by bit, vanload by vanload, we moved everything, including ourselves, to the new house, a cavernous, peach-colored rental with gingerbread trim up in a more urban area of St. Tammony.

One box remained at the house in Piney River Country Club. It was a big box, filled with the things that didn't get packed until the last minute: Becky's brand-new red leather shoes with scalloped edges, a two-liter bottle of root beer, a set of training wheels, a single blade from our Hunter ceiling fan, Mom's old silver mesh evening bag, and our toothbrushes. One Saturday a few weeks later we drove over to pick it up.

"You didn't leave a box here," the woman who now lived in our old house said, flicking her straight brown hair.

"Are you sure?" Mom said. "It was a pretty big box. It had some red leather shoes in it, and a silver mesh purse?"

"Nope, didn't see it," the woman said, starting to close the door, then opening it back up a little. "Oh, wait. I think my husband took that box to the dump. You left it here for three weeks, so we didn't think you wanted it."

"To the dump?" Mom said. "I thought you said you never saw any box."

"Yeah, well, I'm not sure, but yeah, I think he took it to the dump out on Dog Pound Road."

"Why would anyone do a thing like that?" my mother said, stunned. "And after I left those curtains up for you and everything."

"Look, we had to spend a lot of money painting the master bedroom, with all that obscene drawing all over the walls!" the woman said, slamming the door.

We all sat in the van, feeling furious and helpless.

"We can't let her get away with this!" I said, grinding my teeth together as if they were kindling I'd use to set fire to the old house.

"Don't worry, we won't," Mom said, putting the van into reverse and backing out of our former driveway.

Monday morning after we were in school, Mom drove over to our old house, parking around the corner. When she saw that there were no cars in the carport, she opened the back door using her old key. She searched high and low but couldn't find the box.

"That woman's a liar," Mom reported when we got home from school. "They haven't painted over those walls yet!"

We went back to the old house and rapped on the front door. The woman answered it, her hair in a towel and face craggy without makeup.

"Where at the dump did your husband put the box?" Mom asked.

"I don't know!" the woman said. "He just threw it out his truck and left. Look, this isn't my problem."

"You're right, it's our problem, isn't it?" Mom said, savoring every word. "By the way, how much did they charge to paint over the walls?"

The lady's eyes opened wide.

"Have you been inside our house? You're crazy!"

"What size shoe do you wear?" I yelled.

"Sarah, I'll handle this!" Mom said to me.

"I'm calling the police!" the woman said, slamming the door again.

We drove down Dog Pound Road and parked at the entrance to the dump. Acres of matted, moldy garbage spread out before us. We looked around for a few seconds, then got back in the car and drove away.

✲ ✲ ✲

Our new house was an old wooden thing up on cinder-block stilts with a ceiling fan in every room.

"I'm just happy to have my hardwood floors, like we had back home," Mom exclaimed.

"Home" to Mom was still Kansas City, even though we'd lived in Louisiana for almost ten years.

"I seem to remember dark green shag carpet in our house back home," I said.

"And no central air!" Mom continued, exultant. "Look at this screened-in porch out back—I'll put the Hunter ceiling fan from the old house up out here. We've still got three blades for it. That's enough to kick up a breeze."

"Oh yes," I said. "Southeast Louisiana's known for its fresh, stultifying 100 percent humidity breezes."

"Just watch your heads out here," Mom said, ignoring me. "This ceiling's low, and that fan'll come down another eight inches at least."

The rest of the single-story house had fourteen-foot ceilings that made our furniture look like doll furniture. The house was divided down the middle by a wide hall. On one side was a row of four interconnected bedrooms, each with a fireplace fronted by an ornate gas heater. Wedged between the last two bedrooms was the lone bathroom. It had a claw-foot tub and handheld shower nozzle. The other side of the house consisted of a giant living/dining area, a breakfast room, and a kitchen with an ancient gas stove.

Between the kitchen and the breakfast room was a cozy little nook lined with cabinets, which I dubbed the butler's pantry. I polished all our silver—a curvaceous water pitcher, a coffee carafe with warming pedestal, two forks and a knife—and displayed it on a shelf, ready for use.

"Uh-huh," Mom said. "You gonna keep those polished?"

"Yes," I vowed, adopting Meg, my cheerful, can-do *Little Women* personality, to suit our new turn-of-the-century fixer-upper. "It will be my special chore, Marmee."

"Tarnish sets in within twenty-four hours in this clammy environment, that's all I'm saying," Mom said, as if she were talking about gangrene.

She was right. It just wasn't worth the upkeep or the cost of Twinkle Silver Polish, especially when the only other items kept on the shelf were a box of Gerber's Mixed Baby Cereal, which my little brother and sisters and I all ate for breakfast, and one of Skinner's Raisin Bran, which only my mother ate.

"It's an acquired taste," Mom said, poking at the wilted, paper-flavored flakes with her spoon while we gummed our baby cereal.

It was something we'd all gotten used to having around. Baby cereal was quite delicious when crunched up with several teaspoons of granulated sugar. Plus it was high in iron. I could feel my bones crackling to life after I ate a bowl of it.

I couldn't squander my savings on luxe cereals anyway. I had recently shot my wad on a bottle of Fresh Start laundry detergent. Fresh Start was way better than our usual box of soap flakes because, though a traditional powder, it came in a plastic bottle, and it contained sparkling crystals like Folgers coffee.

"This stuff's full of phosphates and causes foaming in rivers and lakes," Mom said, consulting the label.

She refused to let me use it in our washing machine. Fine, I would just use it for my all my fine hand-washables, which I hung up over the claw-foot tub to dry. Our dryer had broken.

"We'll just use a clothesline out in the side yard," Mom said. "It's better for the environment, anyway."

Line-drying clothes loses its romantic luster when it's mandated rather than chosen. Now not only were my clothes gray from being laundered with non-phosphate-containing soap flakes, but they were stiff and crackly from the sun, with rivulets of mildew in the seams.

The rent on the new house was steep, $750 a month. We could afford it for a while. We had some money left over from the sale of the old house, minus what we'd spent on Catholic school tuition. Besides, a BankAmericard with Mom's name on it had come in the mail.

We lived two blocks away from Prompt Succor, the school Becky and I were lying to Dad about not attending. It was slightly embarrassing to be seen walking to school, but less embarrassing than driving up in the Gremlin. It had never been a particularly appealing car to begin with, and it was aging badly. The floor had begun to rust through. I could see little patches of road rushing by my feet when I sat in the passenger seat.

"Rest your feet lightly down there," Mom said to Becky, driving off in her sturdy van without a care in the world.

Who had decided it was a good idea to name a car after a meddlesome, troublemaking imp, anyway? Ours was the only Gremlin I'd ever seen on the road. Everyone else at Prompt Succor drove Mercedes Benzes and BMWs and TR7s, sometimes handed down from their parents, but impeccably maintained. The only other girls with no transportation were Jean Fussell, who couldn't drive because she had a glass eye, and Wanda Fussell (no relation), who couldn't drive because she was a midget.

The first week of school, I decided to run for freshman class president. Nobody knew yet that I shouldn't be class president, least of all me. Nobody knew I was the little sister of the pudgy, impoverished goody-goody in the junior class who made straight A's. No way was I going to get dragged down by Becky. I was chilly when we crossed paths on campus.

I was running against Paige Evans. Paige possessed three things that made the election's outcome uncertain: big boobs, a pukka shell choker, and a cutting-edge, permed-mullet hairstyle. Any one of them could throw the election in either direction.

Lucky for me, Paige came to school on Election Day with a fresh, purple hickey garland across her neck. There were interwoven bite marks, like links on a chain. She had tried to conceal it with cheap foundation. That made it look like a crooked orange grin, with pukka beads in the middle like snaggleteeth.

"What a slut!" everyone said. "I'm totally voting for you, Ya-Ya."

Ya-Ya was my new nickname, given to me by Gigi Wallace. Gigi was a sophomore. She had also gone to Camp Hickory Bluff, no doubt with her father's full knowledge and blessing. She hadn't spoken to me once at camp, but now it was like we were best friends.

Back in Kansas City I'd been known as Sadie Soda-Pop. I was descended from a long line of nicknamed ancestors on my mother's side. Mom's mom was named Isabel, but everybody called her Pat. Grandma Pat had died when I was two months old. At a big family barbeque, she fell backward off some porch steps. Mom tried to grab her, but Grandma slipped from her grasp.

"If I'd just caught her, she wouldn't have died," Mom sometimes said, ignoring the autopsy results that showed a burgeoning aneurysm and a tumor the size of a grapefruit. "You were the last person on Earth she held, Sarah. I think her spirit lives on inside you."

Even if I couldn't have an all-out alias like Grandma Pat, I deserved a nickname. Since we'd moved to Louisiana, no one had cared enough to give me a new one. I'd tried to stir up some interest several times, but Sarah was just about the most boring name you could have and no one was interested in spicing it up.

"Why don't you give me a nickname?" I asked Trey Blanchard in the back of the movie theater the summer before ninth grade.

Trey was a friend of Donna's boyfriend Buck. They'd pimped him off on me and were humping in the back of the theater. Trey had just whirled his tongue around my mouth like a boat propeller.

"Your name reminds me of a crusty old lady out in the desert, trying to have a baby," Trey said, simultaneously showing off his Bible knowledge and making me regret bestowing upon him my first French kiss.

Well, fuck YOU Trey Blanchard and thank *you*, Miss Carson "Gigi" Wallace! It was about time.

Once I was sure Ya-Ya didn't stand for anything negative—I don't think it stood for anything at all—I snuggled into it like a fur coat. I returned the favor by shortening Gigi into "Geege."

I aspired to be Geege's sidekick. She wore horn-rimmed glasses and her constant tan set off her large white incisors so attractively. As a uniform blouse, she wore her father's dress shirts, monogrammed with her own initials. All Geege's clothes were monogrammed, even the slouchy socks spilling out of her L. L. Bean blucher moccasins.

Freshman year, *The Official Preppy Handbook* was my Bible. Nicknames were de rigueur and noblesse oblige. If you didn't have a nickname, you were N.O.K.D. and probably couldn't handle your G&Ts.

Ya-Ya wasn't listed in the *Handbook*'s index of common preppy nicknames, but that was okay. I wasn't sure I could pull off any that were listed, like Muffy or Tipsy.

Mom had transferred that Lacoste alligator so many times, it was looking a little threadbare. I coached her in practicing the Polo pony insignia with the appliqué attachment on her old sewing machine, but pedal power just wasn't going to cut it. At least I had one monogrammed sweater and a tartan kilt.

I adopted a classic hairstyle—a chin-length bob with bangs. My glasses, a pair of beige frames I'd had since fifth grade, could almost pass for tortoiseshell. (At least the arms didn't connect at the bottom of the lenses, like Becky's lavender frames with gold accents.) With such timeless constants forming the base of my look, I could come across as convincingly Ya-Ya.

"Mom, I won the election!" I announced after school that day.

"Darling, that's *wonderful*," she purred in the dead giveaway voice that told me they'd shown *Auntie Mame* on the afternoon movie again.

Mom professed to hate TV, except for when it showed *The Wizard of Oz, The Sound of Music, The Carol Burnett Show*, and *Monty Python's Flying Circus. Auntie Mame* was one of her favorite books, and even though she wasn't exactly fond of Rosalind Russell—nor any other Hollywood actress, because they had all had abortions and bleached their pubes—any time *Auntie Mame* was on, she always watched it. It was to her what *The Incredible Mr. Limpet* was to me.

Our current television was a smallish color affair that sat on top of a large inoperative console TV. Too costly to repair (the tube was blown) yet too functional to get rid of, its sideboard-like dimensions and handsome genuine wood-tone paneling qualified it as a piece of furniture fine enough for the living room. The whole thing sat smack in front of the fireplace, which was rendered purely decorative by the starling colony in its flue.

Auntie Mame was all right, I guess, a little long. I was gratified to learn that rich people were also crazy. My favorite character in the movie was Gloria Upson, the fiancée of Mame's nephew and ward Patrick. A cool blonde, she was all locked mandible and jutting clavicle, finding humor in the fact that "Patrick spoke French to the counterman at Schrafft's, can you *imagine?*"

I had no idea what Schrafft's was. I just liked the idea of nattering along in French while wearing a strapless silk gown the way Gloria did. I could see why Auntie Mame thought she was a phony—I'd read *Catcher in the Rye* at least twenty-three times by then—but *mon Dieu,* wasn't everybody a phony, after a fashion?

"Get packed," Mom said, dropping the Mame accent. "Your father's coming to get you all for the weekend."

"Maaaaa," I began to protest.

"Or you can stay home with me and paint the bedrooms and clear the yard."

I got packed in a hurry. When Mom got into one of her bootstrap modes, it was best to stay out of the way. Blistered hands and a farmer's tan wouldn't impress the counterman at Schrafft's.

Dusk was falling when my father pulled up on the gravel shoulder in front of our house. Mom was already out in the yard chopping down bamboo with a machete.

"What's your mother doing?" Dad asked.

"Clearing that bamboo," I answered. "She says it's full of rats."

Two days later on Sunday, Dad dropped us back off. The yard was stripped down to a lawn and not much else. The house looked naked. With the thick sheath of bamboo gone, you could totally see all our cotton granny panties hanging out on the line. Dryer still broken.

Becky, Hannah, Debbie, Michael, and I walked into the house, which was quiet except for the groaning chorus of the ceiling fans. The aroma of fresh paint was nauseating. We found Mom sitting in the breakfast room, staring a thousand miles across a bowl of congealed Skinner's Raisin Bran.

"Oh, I'm so glad you're home. I don't know if it was the paint fumes or what, but I swear those ceiling fans started talking to me," she said, laughing strangely. "I'm gonna get Bubba Jenkins to come finish up the yard. It's just too much work; I got all scratched up out there."

That night I awoke with a start.

A dark shadow flitted across the bottom of my bed. I tried to call out "Mom!" but my throat felt too tight to speak. My ears roared like two vacuum cleaner nozzles were hooked on them. My entire body felt locked up in rictus—*had I stepped on a rusty nail recently?*

The only part of me that could move was my eyes. The clock said 4:37. Wasn't that what they called the Hour of the Wolf, when most murder and mayhem were committed? Where had I read that? *Helter Skelter?*

My bedroom was in the front of the house, with doors out onto the porch. Doors held shut with flimsy hook-and-eye locks. The lady who lived next door told me this house used to belong to an undertaker and his family. My bedroom had been his office: where the bodies were embalmed.

The shadow flitted back across the room. I heard soft hoof steps. This was no ordinary dead person. This was the Devil.

I lay there, waiting to die of fear, like people who jump off buildings do in midair. Finally, the suction in my ears died down and my neck tendons loosened up. I got some feeling back in my legs and swung them over the side of my bed. I ran through Becky's room and into my mom's.

"Wha? Whatsit, is it? Sarah?" she said as I climbed into bed with her.

"I think I just saw a ghost in my room," I choked out.

"Yeah," Mom said, shifting over to one side of the bed to give me more room. "I woke up Saturday night when y'all were gone at your father's and saw a grinning head floating over the end of my bed, like it was looking over his shoulder at me. And then the ceiling fan started saying, 'Get out, get out, get out' and I thought I was going to die."

"Mom!" I cried. "Don't tell me things like that!"

"Well, I just thought it would make you feel better," she said, patting me.

I suppose there *was* an Amityville Horror kind of glamour to it all. Perhaps there could be a certain cachet to living in a haunted house.

"Yeah, there were flies all over our sewing room," I would tell the girls at school. "There'd probably be blood and dead bodies in the basement, if we had a basement, but nobody has a basement in Louisiana but we would if anybody did."

I drifted off to sleep next to my mother, the ceiling fan drawling above us.

Yalp yalp yalp.

✲ ✲ ✲

"DING—"

"I'll get it!" I yelled, running toward the front door and then stopping short to compose myself.

"DONG!"

Through the sheer lace curtains I could see a manly outline. I opened the door, smiling.

"Hello?" I said, shielding my eyes and trying to make out his backlit features.

"Ya mama home?" scritched a voice of indeterminate age.

"Well, hey, Bubba!" Mom said, coming from behind me and stepping out onto the porch. "Let me show you the yard."

I watched out the window of my room as Mom showed a guy wearing a baseball hat around our yard, pointing to the last straggling stand of bamboo and making a hacking motion with her arm. She laughed, they shook hands, he spat on the ground and walked over to a pickup truck that had a lawn mower and rakes in the back of it.

"Who's that?" I asked when she came back in the house.

"That's Bubba Jenkins. He does yard work. He only charges twenty dollars," she answered, peeping out at him through the curtains.

We had a manservant!

I joined Mom at the window. A manservant who was peeing on the bamboo, but a manservant nonetheless.

"Good Lord, did you see his filthy hat?" Mom said. "I certainly hope he doesn't wear that all the time. I'd hate for someone to drive by and see him wearing that hat and think I agree with that sort of thing."

"Mother, of course his hat's dirty," I said, all tight-jawed. "The poor man does yard work, for heaven's sake."

"No, not dirty like that," she said, lowering her voice in case Bubba could hear over the lawn mower and through the walls. "Didn't you see it? It says right on the front there in big red letters, 'I'm on a *P-U-S-S-Y* hunt.'"

✵ ✵ ✵

"I just wish you'd asked me first, before you went ahead and asked someone," Mom said.

"You would've just said no!" I said, clanking my fork down so hard I chipped the plate.

We were sitting around the dinner table eating 25-cent generic chicken potpies from Delchamps. They came in plain white boxes with big black block letters that said "CHICK PIE" on them. When you gave up a name brand, I guess you had to lose a syllable too. They were delicious; we ate them about three times a week. If we had guests, I would tell them we had once bought some pricey Swanson Pot Pies just for the little tins, so we could use them to make our own special homemade potpies from a recipe handed down from my great-grandmother. I still hadn't figured out how to explain the Hershey's Cocoa tins in the corner of every room. They were my mother's invention: humane, chemical-free roach traps. Roaches climbed in, but they couldn't crawl out, though they would skitter about desperately for several days.

"Why do you want to take Mickey Foote to a dance anyway?" Mom continued. "He has little green rat teeth."

I didn't know why I'd invited Mickey Foote to the Mock Homecoming dance. He did have little green rat teeth, and they flickered with disappointment when he accepted my offer last night after the football game at St. Jude's.

"Yeahhhhh, I guess I'll go with you," Mickey said, looking around the milling crowd exiting the bleachers for something better to come along.

Being class president meant a lot less than I'd hoped. No one had asked me to Homecoming. Mock Homecoming was a dance where the girl could ask the guy. Of course I wanted to go.

"Mickey Foote is nice," I said weakly.

"Yuck," Becky said, stabbing at the lone gray pea in her chicken potpie. "His sister Charlene's in my class and she still eats her boogers."

"Becky! Their mother works hard," Mom said. "Poor thing, her husband's got a drinking problem."

"Maybe all they have is boogers to eat," Hannah guffawed.

Debbie said nothing. She was trying to put some distance between herself and any association with booger eating.

"How much are we talking here?" Mom said.

"I have to buy the tickets to the dance. They're $25. I have to get him a boutonniere," I said, slowly warming to the real bombshell. "And then of course, there's dinner beforehand."

"Accchhh! I suppose you'll have to go to Ching Chong along with all the other ritzies," Mom griped.

"It's *Chung Chow*," I said, disgusted.

"Whatever," Mom said. "That'll be another twenty bucks or so."

Hardly. Forty bucks, minimum, and that's if I didn't eat anything myself, and what would be the fun of that?

"Look here, I saw in the *News Banner* there's a Valley Girl contest down at Bogue Falaya Mall tomorrow," Mom said. "The first prize is fifty bucks. I bet you could win it, Sarah."

"Hey!" Michael said from his makeshift booster seat, a stack of encyclopedias on the chair at the head of the table. He was wearing Debbie's curly red Little Orphan Annie wig and a green tutu. He had both hands down his training pants and a look of sheer delight on his two-year-old face. "There's a little bag down here with balls in it, and when I pinch them, it hurts!"

"Eat your supper," Mom said, taking a bite and winking at us, as if food were a conspiracy.

The next morning I put on Becky's white jersey miniskirt and my turquoise blouse with the asymmetrical button placket. I gave Frank Zappa's "Valley Girl" 45 a few listens on the turntable. Feeling a little tired, I took four NoDoz caffeine pills.

Bogue Falaya Mall was a cross-shaped building with an atrium in the center. At one time it may have been a flourishing center of commerce, but now all that remained was the five-and-dime Murphy's, a rarely fashionable clothing store named Bealls, and a numismatics shop run by a cranky old man with a beard. Once I'd tried to sell him an old stamp commemorating the Apollo-Soyuz space linkup.

"See this postmark in the corner?" he'd said.

"I see a tiny speck, if that's what you mean," I answered.

"Ain't worth the eight cents somebody once paid for it."

In the middle of the atrium was a large square planter filled with scrubby ferns. A plank of plywood had been laid over it to serve as a stage.

I had two competitors: a chubby eight-year-old whose cuteness factor was diminished by her exposed cascading belly rolls, and another girl around my age who was wearing overalls with nothing underneath. Both of them were obviously in it for the money and didn't even know what a Valley Girl was. Freeloaders.

The audience consisted of my mom and siblings, the two other contestants' mothers, two old men playing chess while another old man watched, a skeptical-looking coin collector who stopped by on his way to the numismatics shop, and ten assorted old ladies in matching paper sailor hats. All of them wore shoes that looked like baked potatoes.

The other two girls stank. I went last, delivering a zippy, extemporaneous monologue about my "totally bitchen" boyfriend who drove me in his "BMW" to "Ventura Boulevard" to buy some "rilly, rilly" cute clothes, "fer sure."

I had a good feel for my audience. When the old ladies started unwrapping hard candies, I sensed it was time to wrap it up.

"So you know, like totally? And what-EVER. I mean, totally awesome. For sure, rilly, y'know, like, for sure!" I trailed off, bowing and jumping off the stage.

"Well, let's hear it for all our Valley Girls," said the emcee, an assistant manager at Bealls. "You were all wonderful, but only one of you can win the fifty dollars. And that would be Sarah . . . Thay-er? Tire? Ty-ree?"

"It's THIGH-er," I said, leaping back onstage and grabbing the check while my family applauded wildly.

The chubby girl cried. The girl in the overalls kicked one of the old ladies in the sailor hats in the shin, which made the chubby girl stop crying and laugh.

My sisters gathered around me to look at the check.

"Fifty dollars," I said, flicking it.

"Wait a minute," my mom said, looking like she might bite the check the way people in cartoons bit coins. "That's not a fifty-dollar check. That's a fifty-dollar gift certificate. To Bealls."

"Well, I'll just buy something small and get the rest in cash," I said.

I went over to Bealls' cosmetics counter and bought a small package of lily of the valley–scented bath cubes. You could use them as drawer sachets until they started to crumble, then throw them in the tub. I held my hand out for the $47 in change.

"Sorry, store credit only with a gift certificate," said the counterman.

I didn't waste any French on *him.*

"Might as well spend it all," Mom said. "You could invite your sisters to each pick something out."

I didn't hear that last part, I was already in the dressing room pulling on a matching Esprit top and pair of culottes. At least I would have something to *wear* to the dance.

"Since we're already over here, let's drop by Gibson's," Mom said. "They've got Moonlight Madness going on tonight."

"Can I get an Icee and popcorn?" Debbie and Hannah asked in robotic unison.

"Maybe," Mom sighed. "We need toilet paper."

Gibson's was across the street from the mall. It was sort of a precursor to Wal-Mart, a sprawling warehouse-type store. Gibson's carried all the staples. It also had a special section filled with racks and bins of fire-damaged merchandise.

"Some days I've hit on a real bargain in there," Mom said, a dubious brag.

I was a little nervous about going to Gibson's. Last week I'd been in wearing Becky's jean vest. Its entire surface, front and back, was covered with pockets, perfect for shoplifting small items, like little boxes of No-Doz caffeine pills. I hadn't been caught, but I doubted it was a good idea to return to the scene of a crime so fresh.

Hannah dove into a cardboard box in the fire-sale section, surfacing with a pair of four-inch, wooden-wedge sandals.

"There's a chip in the heel, but they're only 50 cents," she said, clunking up and down the aisles in them.

"Those are a little old for you, but they look like genuine oak . . ." Mom said. "I guess they're okay for dress-up. You can fill that chip in with plastic wood when we get home." She tossed the heels into our shopping cart and held a singed sweater under Debbie's chin. "This looks kinda snazzy . . . hey look, they're doing portraits today. Hannah and Debbie, let's get your picture taken. You just got haircuts. Perfect timing."

"What about the money I need for Mock Homecoming?" I said desperately, looking at the sign that said *Special Portrait Package: $8.99*. "It's next weekend—we can't afford a new portrait right now!"

"Look missy, you got your Bealls shopping spree," Mom said, signing up on the photographer's waiting list.

She dragged Hannah and Debbie over to the girls' department. She dressed them in matching blouses and sweater vests—"The picture's only from the waist up"—tucked in the price tags, and marched them back to the photographer just as he called, "Thay-er? Three? Treer?"

"It's Thyre," Mom said, slapping nine dollars into his palm.

Hannah and Debbie stood between a wagon wheel and a grey-carpeted mound in front of the backdrop, which was itself a giant photo of an autumn day on a country lane. Their cropped hair and sweater vests lent them a dyke-y air.

"What pretty . . . handsome . . . girls?" the photographer said.

Hannah and Debbie gritted their teeth. The flashbulbs popped three times.

"Okay," Mom said. "Now take off those clothes and hang 'em back up on the rack—*neatly*. I'm going to check out. I'll meet you out front."

"How am I going to pay for dinner at Chung Chow?" I whined, following her.

"Don't be such a molly child," Mom said.

She stopped in front of a display of luggage. She fingered the price tag hanging from the handle of a large faux Samsonite.

"One-O-nine ninety-nine," Mom muttered, resting her fingernails on her lips and giving that thousand-mile Okie stare.

She grabbed the suitcase and dragged it up to the checkout counter.

"Are we going on a trip?" I asked, utterly diverted and thrilled.

"Maybe!" Mom said, winking at the cashier and whipping out the BankAmericard.

Two days later, Mom and I got in the car and drove back to Gibson's. She sent me to the customer service desk with the suitcase and the receipt.

"Sir, my mom bought this suitcase the other day, but when we got home our neighbor had given us one that was the same size and practically brand new, and a real Samsonite to boot," I said. "So you see, we shan't be needing this one."

This was the Stone Age. There was no swiping of credit cards and automatic electronic debiting. One's credit card was placed under a triplicate bill of sale in a little manual embossing device. An inked roller was *kuh-shunked* over the receipt. One copy went to the customer, one went to the store's bank, and the third went to the credit card company. Charges took weeks, if not months, to go through. And when you returned something you'd bought with a credit card, they generally gave you cash back.

"You want your refund in tens or twunnies?" asked the counterlady.

"Twen-TEES, please," I enunciated for her benefit.

Mom gave me sixty bucks for dinner. Mickey Foote and I sat at a table next to the clattering kitchen. He slavered over Chung Chow's menu like an old man panning for gold.

"All right! I'll take an order of egg rolls—that's two, right?—some Crispy Cream Cheese Wontons, Egg Drop Soup, Four Happiness—that got shrimp in it, right?—a virgin Singapore Sling, and also a Coke," he rattled off. "Oh and how about some Rainbow Beef too? That sounds good."

"Mickey, Rainbow Beef is full of onions and comes on these crunchy rice noodles that taste like Styrofoam," I began a bit sharply, but caught myself. "I mean, it's totally like, rilly gross."

"Sounds good to me. I want it," he said, his eyes darting.

Rat-like.

"For you?" our waitress said, stripped to the level of service commensurate with her anticipated tip.

"Wellllll," I said, tracing my finger down the menu. "Moo Goo Gai Pan is my usual, with a side of Special Fried Rice. You do that so well here. And Shrimp Kew is another favorite. . . . "

I was stalling, skimming through the menu and tallying all the prices from Mickey's order on a piece of paper in my head. He'd gobbled up almost the entire sixty bucks!

"You know, I'm not really hungry tonight," I said, closing the menu gracefully. "I'll just drink this water. I'm on a diet, anyway."

Later, at the dance, Mickey and I had our picture taken.

"Smile," said the photographer, thankfully not the same one from Gibson's.

Mickey bared his little green rat teeth, inlaid with shreds of barbequed pork and wonton. I pressed my lips tightly together and clung to his arm, trying not to faint dead away of hunger.

✵ ✵ ✵

On Christmas morning my dad came over to deliver presents. We huddled in Becky's bedroom. There was no central heat in the house, and Mom didn't like using all the gas space heaters because they were "full of asbestos." During the cold months, we pretty much lived out of the two middle bedrooms.

I got the jambox I'd asked for, and what I thought was a sky blue Lacoste shirt but turned out to be Le Tigre—ugh.

"Thanks," I said flatly to my father. "It's rilly GREAT."

"And here's something to go along with that shirt," he said, handing me a small rectangular package that obviously contained the worst class of gift ever: the dreaded BOOK.

I unwrapped it.

"Oh yeah, WOW!" I said.

This was more like it—a deluxe hardcover version of *The Official Preppy Handbook,* complete with a gold-embossed, rich maroon fabric binding sleeve. I only wished Dad had taken a peek inside before buying the Le Tigre shirt. He stayed for a couple of hours. Mom stacked the two dozen donuts he'd brought on plates. We munched on those throughout the morning, tinkering with our new belongings.

As the day progressed, it grew colder. Dad left before Christmas dinner, which we traditionally sat down for at two. We ate the turkey on Becky's bed.

"Great, you got mashed potatoes and gravy all over my bedspread," she complained to Debbie.

"That's not the Christmas spirit," Mom said cheerfully.

Later that evening, I huddled next to the space heater wearing my mom's old down-filled nylon jacket, examining the seams around the

Le Tigre insignia to see if it could be easily removed and replaced with a ratty, transplanted alligator. Stupid Dad, thinking Le Tigre was preppy.

Soooo cold.

I scooted closer to the heater, and popped a tape into my new jambox.

"Meli keliki maka is the thing to say / On a bright Hawaiian Christmas Day," sang Bing Crosby and the Andrews Sisters.

I began to warm up. What did it matter if Dad got me a Le Tigre shirt? He meant well. It was the thought that counted. He was just trying to make me happy, wasn't he?

My cold heart thawed a little. I shouldn't be such a Scrooge.

I shivered.

I felt Christmasy. A light snow began to fall around me, coating my shoulders and the floor. The smell of a cozy fire in the hearth crept into the air, then bit by bit turned acrid.

"Aaaaaaaaagggghhhh, HELP!" I yelled. "I'm on FIRE!"

Everyone came running in from the other warm room, Mom's bedroom next door. I'd skootched too close to the heater. The back of the down jacket had melted and feathers were flying through the air, some glowing, ignited by the space heater.

"Sarah, HOLD STILL," Mom screamed. "Roll on the ground, roll around! Wait, that's nylon. It'll melt onto your skin and give you third-degree burns. I think I'm supposed to throw water on it."

"No, throw flour on it!" Becky said. "That's what you're supposed to do with grease fires."

"No, you're supposed to suffocate it—fires need oxygen," Mom said, grabbing the bedspread off Becky's bed and swinging it at me.

I fell to the floor on my back, releasing a final puff of smoldering feathers.

"Great," Becky said. "My bedspread is officially ruined."

✲ ✲ ✲

Finally, the weather got warmer. One night we ate dinner out on the back porch, under the shuddering three-bladed ceiling fan. It was

Sunday and Lent would start on Wednesday. A long hard winter of chill and deprivation had put me in no mood to give something up.

"I don't see the point of giving something up for Lent," I said.

"Sarah, you will too," Mom said, putting down a platter of pancakes and a bottle of genuine 100 percent pure maple syrup. "It's a small sacrifice, compared with what Christ gave up for us."

"Why can't we have Log Cabin?" Becky said.

"Because it's ninety-five percent corn syrup," Mom said.

"I'm giving up gum for Lent," Hannah said.

"You hate gum!" I said. "That would be like me giving up mayonnaise—hey, I'm giving up mayonnaise!"

"You should have to give up mustard," Becky said.

"Okay, then you give up pickles," I said, referring to her jar-a-day habit, juice and all.

"I have an idea," Mom said. "Let's give up arguing at dinnertime and remember to say grace every night. Thank you, O Lord, for this food."

We ate quietly for a few minutes. Michael sat on Mom's lap, nursing.

"Ecch, you're such a hippie," I said. "When are you gonna quit that? He's almost three."

"Mind your own business!" Mom said, lifting her other boob to squirt milk in my direction.

"GROSS!" we all squealed, our mouths full of pancake.

When the platter was empty, Mom got up to clear the table.

"Becky, take him," she said, handing Michael over.

"Wheeeeeeeeeeeeeeeeee," Becky said, spinning him around and around, then throwing him up in the air. "You're an airplane! You're an airplane! You're—"

GANK!

A blade of the ceiling fan clonked into Michael's head about an inch back from his white-blond hairline.

There was a simultaneous collective sharp intake of breath. Right when we started to exhale, blood poured out of the gash.

"Shit!" Mom said. "Becky, pull the car around!"

Michael's mouth released a long, mewling sound.

We all froze. Michael was a pretty even-tempered little kid, not prone to tantrums or extreme emotional displays. When he was one, he had fallen down and cracked the enamel of his four top front teeth. Now they were capped in silver. Seeing him hysterical was especially frightening. He looked like something I'd always believed impossible: a robot capable of human emotion, human *baby* emotion.

Mom grabbed Michael and wrapped his head up with a thin white cotton dish towel. Blood soaked it immediately. Becky pulled the van up to the back of the house and Mom ran down the steps from the porch and got in the passenger seat, clutching Michael.

Hannah and Debbie and I were all crying hysterically. I ran to the phone and called St. Tammany Hospital.

"St. Tammany Hospital help you?"

"Yes, my mother's coming over there in a green-and-white van with my brother his head's bleeding hit on the ceiling fan so be on the lookout for them!" I couldn't stop sobbing.

"Okay, ma'am," she said. "Calm down, ma'am. We'll take good care of him."

I hung up. Exactly a minute later, I picked up the phone to call them again.

The line was dead. That month, Mom hadn't been able to pay the bill.

* * *

"He was such a little angel," cooed Mom, bouncing Michael on her lap. "Halfway to the hospital, he sat up and said, 'It's okay, Mom. I'm all right!' They strapped him into a papoose-looking thingy and gave him three shots in the head and stitched him up and he didn't even cry. They said they didn't think he has a concussion, but we should keep him awake for a while just to be on the safe side."

Michael's big eyes rolled slowly in his head, a once-cute gesture that sent a spasm of fear through me.

Please God. I'll give up mayonnaise for Lent.

No wait. Sorry, God. I mean mustard.

Two days later was Fat Tuesday, Mardi Gras. We always went across the lake to the Argus parade in Metairie, the closest suburb of New Orleans. The Krewe of Argus, the social group that put on the parade, was all-female. They threw the usual plastic beads and multicolored doubloons, and if you were lucky you might catch one of their plastic cups. Beads were only worth it if they were the opalescent, ping-pong-ball-sized variety. Cups were even better.

Personally, I had my sights set on something deluxe. This was the year I was gonna get one of those feathered rubber spears or a giant novelty toothbrush. Those were the two most coveted items, aside from a golden coconut from the Krewe of Zulu, an all-Black krewe whose parade rolled its way down St. Charles Avenue through the ritziest, most historical neighborhood outside the French Quarter. I didn't waste my time pining for a golden coconut; I heard they only threw those to other black people or rich white people.

At the parade, I hoisted Michael up on my shoulders and squeezed my way as close to the front of the crowd as I could. In addition to the jagged Frankenline of stitches across his scalp, he had a smudge of dirt on one cheek that looked like a sad bruise. I relaxed my face, letting my lips fall downward into a forlorn rainbow. I drew up some worry lines between my eyebrows.

"Hey, throw me somethin'! Hey!" I called, loud yet melancholy.

Once the ladies of the Krewe of Argus noticed us, they pelted us nonstop with entire plastic bags of beads and stacks of cups. It was all I could do to shove everything into the paper grocery bag clamped between my legs while keeping my face upturned and needy. Doubloons tinkled on the pavement around us.

"Those're ours!" I said, desperately gouging them from the hands of children and the elderly. "They threw those to *us*!"

Our bag was brimming over, bursting open. The last float of Argus was rolling by.

"Please ma'am, PLEASE," I yelled meekly at a masked lady above. "My son here would like a spear. Or a giant toothbrush. Or both, if you can spare them."

The lady nudged another lady, who turned around and dug in a box beneath a large, grinning, bobbling, papier-mâché devil head. She turned around with the spear and the toothbrush in her hands. The float kept rumbling along—she was getting away!

"ME!! It's for me," I cried, running alongside the float with one arm outstretched and the other clamping Michael's legs together at the ankles, to keep him on my shoulders.

Everyone around us began reaching upward with their filthy paws.

The lady on the float seemed troubled behind her devil mask. She was trying to reach over everyone's head to hand us the glorious spear, the sleek plastic giant toothbrush. Right when she shrugged and seemed ready to give up, I pushed a gaggle of children out of the way and grabbed the end of the spear.

"Michael, get the toothbrush!" I said, pinching his leg.

He got it! We got them!

Compared to our big bag o' booty, my sisters had very little to show for themselves. On the way back to the van, everyone took a turn chucking the spear and brushing their teeth. I couldn't believe our good luck.

We went through the Burger King drive-thru before getting on the Causeway to go back over Lake Pontchartrain.

Sleepy in the back bench seat of the van, I bit into my cheeseburger. I'd had it my way, all right: extra mustard dripped down my chin.

Oh lay off, God. I'm giving it up tomorrow.

Misty Popularity

"Whatchall think of Dionne's work here?" said Philip "Tootsy" Broussard, Prompt Succor's art teacher. He was known locally for his oils of the sun setting over cypress trees in the swamp.

Tootsy clipped a piece of paper onto the easel. It was a pencil drawing of a large lone cone, shimmering on the horizon.

Dionne Domanqué was the star of Tootsy's class. She and I had been best friends for a month back in seventh grade. Every day after school we played Escape from the Nazis, following the same never-changing script. Dionne's bedroom was Berlin, where the two of us ran a hybrid orphanage/charm school together. After applying broad strokes of her mother's rouge across our cheekbones, we would make our one orphan, her little sister Camille, practice an intricate tap-dance routine over and over, threatening to rap her knuckles with a dowel rod if she flubbed it up. Camille would then perform as if she were on the *Gong Show*, with Dionne as the celebrity judge Jaye P. Morgan, and me as Rip Taylor. We were merciless, gonging her from the get-go. Camille kept dancing and crying until the imaginary Nazis burst into the room. As the Nazis dragged Camille away to Treblinka, Dionne and I would run into the closet and lie very still on top of each other, arguing about how we were going to pay the bills. Our orphanage went out of business, and shortly thereafter, so did our friendship.

"Come on, y'all," Tootsy said, waiting.

I raised my hand.

"I think it really captures the encroachment of the machine age upon our barren, like, post-nuclear landscape," I said, beaming at Dionne.

Dionne looked at her lap and smiled humbly. She designed and executed all the murals and photo backdrops for our school dances. An Evening in the Vieux Carré, Venice at Dusk, Poppin' Frosh, Luvs Baby Sophomores: you name the theme, she could flesh it out with buckets of tempera paint.

"Well, I don't know about alla *that*," Tootsy said, removing Dionne's pencil drawing from the easel and replacing it with one of his own paintings. "Now let's alla y'all take a crack at this bayou scene."

I fastened a medium nib to my quill and dipped it into black India ink. My long-range plan was to move to Paris and become an artist. I would have an apartment in the Eiffel Tower and a simple pied-à-terre in the Latin Quarter, where I assumed they spoke Español. (I was taking Spanish because Becky had warned me that Prompt Succor's French teacher threw chalk.)

I glanced at Dionne. Her hands fluttered lightly over her sketchpad. A perfect reproduction of Tootsy's swamp materialized from all areas at once, as though emerging from a magical mist.

Of course it did. Dionne's father's watercolors of New Orleans street scenes hung in our local McDonald's. She was probably doing two-point perspective and three-dimensional shading in her playpen. Both my parents could draw well but had never taken the time to nurture my innate talent.

I stared at my own work thus far: a single Y-shaped tree trunk rendered in thick fat lines. My hand around the quill felt like a giant meatball on a stick. The sorrowful sword of self-knowledge pierced my heart: I would never be an artist. I drew the outline of everything before coloring it in. Even in coloring books, where the outline was already provided, I felt the need to retrace it with whatever color crayon I intended to use to fill it in.

I was still coloring in coloring books.

This compulsion to outline extended to my schoolwork, the grocery lists I was trying to teach my mother to keep, and my Duran Duran fan fiction hobby. I kept lists obsessively, including one master list entitled "Lists." Life was tidier, more orderly with lists.

I just happened to be very word-oriented. When I thought of an object, I didn't see the object itself, I saw the word. When I thought of God, the word "God" appeared in my mind's eye in bold Gothic relief, sometimes but not always backlit by the sun's rays streaming from behind a cloud. Words and fonts: they were my forte.

At the bottom of my swamp scene, I signed my name in capital Art Deco–style letters.

✵ ✵ ✵

"Ucch, I *hate* reading!" Marzy Aubert said to me in the locker room during lunch period.

The crust around her eyes crinkled in disgust. She had some kind of allergy that made her eyelids flake and peel.

"Huh?!" I said, startled.

I tossed *A Tale of Two Cities* to the ground in a convincingly reflexive, hot-potato motion. We'd already read it in English class, but I was rereading it just for the sheer pleasure. The machinations of Madame DeFarge were evil to be twice savored.

"Reading?" I said. "Gross!"

One had to tread delicately about such matters: better safe than sorry. Winning the offices of Student Council treasurer and Spanish Club historian had secured my political career, but maybe I shouldn't have gone for Literary Club secretary and Library Club parliamentarian. I didn't know if Marzy was an ignorant anomaly, or if hating to read was the latest acid test for popularity, something that always seemed to be just out of reach.

Lord knows I'd tried everything, run the gamut. In addition to Yoda's entire "Jedi you are" speech, I could now imitate Belinda Carlisle from the Go-Gos singing "Vacation," *and* the vaguely Christian cartoon character Jot *and* that roll of talking toilet paper from the TV commercial.

"Hey-I'm-new-Banner-tissue-and-yer-gonna-LO-O-OVE-me!"

Still, soul mates—or even better, minions—had eluded me.

Fine. I would pretend to hate reading to be popular, but I couldn't give it up. *Nay!* Reading enabled me to see that there was nothing wrong with eating fried eggs for dinner every night. It was something Huck Finn would do, if he were lucky enough to steal some from a chicken coop. Stolen eggs, fresh from the coop—they taste so good, Jim.

Reading elevated an everyday bed to a sublime "pallet." Since the TV on top of the broken TV had broken, every weekday between school and dinner, I would retire to my quarters and, rigid with a delectable wistfulness, lie upon my pallet, reading.

One afternoon as I was rereading *The Lord of the Rings, Book II: The Two Towers* for the fourth time, I heard a loud scraping noise coming from our public rooms.

Hark! Be ye the drums of war?

"What's going on out there?" I called, struck with just enough panic to disable Mine Olde English speech mechanism.

No one answered. The sound continued. I crept out to the living room to see a purple-faced man dragging our upright piano from the dining room through the living room, leaving a long swath of scratch marks across the wooden floor.

"I'm selling the piano," Mom yelled over the racket.

"What?" I cried. "I was going to be a musician!"

"You haven't taken lessons since Kansas City," she scolded. "When you went to Mrs. Lion's house, you used to bang your head on the keys."

Mrs. Lion had been my piano teacher. Just the thought of her licking the tip of her splintery, hand-sharpened pencil still made me shudder.

"Well, what about Becky and Hannah?" I demanded.

Both of them were legitimately proficient, willing pianists. They actually saved up money to purchase sheet music from Wurlein's over in New Orleans. Becky's specialty was "Down Under" by Men at Work, and Hannah's was Survivor's "Eye of the Tiger." We were going to put together a cabaret act with me as the chanteuse. Between the two of them, I demonstrated an impressive stylistic range.

"They never practice," Mom said, which I translated to mean *We need the money.*

The man pulled the piano down the steps of the porch—bam, bam, crash, BLONK. He started to jerry-rig a ramp up to the bed of his pickup with some two-by-fours.

"You want me to haul away that ol' AMC while I'm at it?" he said, gesturing toward the Gremlin, which was sagging like a cripple on the side of the house. "I could bring it by the scrap yard for ya, split whatever they give us. I got tow chains with me."

"I tell you what," Mom said, her eyes shimmering with opportunity about to be blown. "Since you're so kind to move that big piano all by yourself, why don't I pay you fifty dollars to take both of them out of my hair?"

Huh? My mouth strained against its frozen smile.

The man tugged his ear as if he was trying to shake something loose.

"You want to pay ME?" he said. "To take 'em both?"

"Oh, *yes*," Mom said rhapsodically, completely out of touch with our dire economic straits. "You don't even know what a help you've been to me today!"

The man drove off with the piano in his truck, towing the Gremlin behind, singing along to Charlie Daniels on the radio: "Cuz I'm still in Saigon / In my mi-i-ind."

I kicked some gravel and turned to Mom.

"I was *too* going to be a musician."

"I had to get rid of it," she said. "We're moving at the end of the summer when the lease is up."

"Why?" I asked, barely able to contain my excitement. Maybe our new home would have a pool or, at the very least, a fountain.

"We can't afford this place anymore," Mom said. "The landlady wants to raise the rent to nine hundred a month."

"What about the money left over from the house on Tchefuncte?" I asked.

"It's long gone," Mom sighed. "I can't stretch your dad's monthly check to cover this place and keep you all fed as it is. I've got a new job starting today."

Job? Ugh. That meant one thing: babysitting. More children. That was the last thing we needed around here. Mom charged so little she ended up babysitting for kids who were as poor as us, and they ate us out of house and home. Any money she made watching them went right back into their mouths.

"Don't tell me you're gonna watch that kid Nick again!" I wailed.

Nick could put away half a gallon of milk a day. He was like that little mouse Nibbles from the Tom and Jerry cartoons. One minute he's on the doorstep in a diaper, all innocent; the next minute he's unhinging his jaw to swallow a roast turkey whole.

"No, Nick's not coming back," Mom said. "His dad doesn't like me."

Last time Nick came over, he peed and pooed his way through all his clothes. When his dad came to pick him up, Nick was standing out in our yard, wearing nothing but a little white slip embroidered with flowers, something Great-Grandma Gertie had made for me when I was a baby. Nick's dad threw him into their car and drove away as if a tiger were after them.

"I got something new lined up, three public school kids, the Theriots," Mom said. "One's a boy around your age, a little younger."

I went back to my undertaker's-office-cum-bedroom and changed out of my school uniform and into my uniform for seduction: a yellow tank top and cutoffs. Except for the pegged, pinstriped Guess? Jeans I'd saved my own babysitting money to buy, all my jeans quickly ended up as cutoffs. Dad would only buy flared bootleg no-name jeans from Tally's Feed-N-Seed. He insisted there was a difference between *want* and *need.*

"When I was a kid, I had two shirts and one pair of pants," he often said. "And sometimes all we had to eat was a piece of bread with gravy on it."

"Well, that gravy had to come from somewhere," I once had the balls to retort.

"Do what now?" Dad had said.

"I mean, you must have had *some* meat, somewhere along the line, to get the gravy from, right?" I'd pushed, cringing at my own mouthiness.

"Don't give me none of your lip, girl," Dad said, letting me off easy, considering.

Jay Theriot and his brothers were well-scrubbed and healthy-looking. Jay's hair was cut into the shape of a candy dish. I sashayed past him, thrusting out my little nickel-nipple niblets as far as they would go.

"You like snakes?" he whispered.

"Who me? Oh yeah, all the time," I said. "Not as much as last year, when I was your age."

We spent the next hour stalking a grass snake in the backyard. I huddled close to Jay behind an azalea bush, its luscious purple flowers in full bloom.

"Mrrrrrrawwwwrrrrr," came a sound from behind us, under the house.

"What's that?" Jay said.

"I don't know," I answered, profoundly embarrassed.

I picked up a few pebbles and pitched them under the house.

"MrrrAWWWRRerrrrrrAWWWWrrr," the sound continued.

Hannah came running down the porch steps.

"Is that Sylvester, doing it under the house again?" she said, leaning over and peering into the dark space.

"Who's doing what?" Jay said.

"NOTHING," I said, more to Hannah than to him.

I took the wad of gum out of my mouth and motioned as if I were going to stick it on her arm.

"Agh!" Hannah said, jumping away.

Mom came down the porch steps.

"The sex fiend's at it again, huh?" she said, shaking her head.

Like our old cat Grandma before him, we had mistaken Sylvester for a male and named him accordingly. Until we saw "him" under the house, having sex with a notoriously horny neighborhood tom. Oddly, they were doing it missionary-style: Sylvester beneath him with her legs spread and front paws circling his neck. Seeing cats so humanly in flagrante delicto made the whole affair seem even dirtier.

Sylvester's love nest was directly under my bedroom. I'd had plenty of sleepless nights listening to cat orgies six feet beneath me.

Debbie and Michael and Jay's two little brothers crowded in behind the bush with me and Jay, milling around.

"Murrr murr murrrr MURWAHHHHHH," Sylvester and her companion cried out in a shuddering crescendo.

"Sylvester, come out here right now!" Hannah called, shining Michael's flashlight under the house.

Out she slunk, looking pretty satisfied with herself. She sneered at us and licked her paw, pretty snobby for a cat who got fed the cheapest cat food on the market: Kozy Kitten. Its label had a cartoon cat wearing a pearl necklace and a diamond tiara, more befitting the kind of cat that ate Fancy Feast. Though knowing Sylvester, a pearl necklace wasn't totally out of the question.

The smartest move would be to pretend none of this had ever happened.

"Now!" I said brightly to Jay. "Where'd that snake go?"

✵ ✵ ✵

Mom's low, low prices had driven down the babysitting rates in our neighborhood. I couldn't make a living if I tried. The whole point of babysitting was going over to somebody else's house, being privy to a brand-new world and its attendant foibles, secrets, jellies, creams, and most important: snacks.

Having other people's children constantly underfoot made me itch to get out of the house. I began to hold children in contempt. They were so needy. As soon as I turned fifteen the summer after freshman year, I went down to the parish courthouse to apply for a work permit. I plastered on a wan expression and held Michael on my hip, thinking everyone would let a young single mother like me cut to the front of the line.

Work permit in hand, I headed straight to Bunz out on Highway 190. Becky already worked there and had put in a good word for me.

Bunz was a new branch of a chain known as THE place to sate your craving for burgers topped with chili, cheese, and what looked to me like a foaming mayo-mustard hybrid. This culinary nadir was available in two sizes: the Big'Un and the Littl'Un. If you ate in, you could get your root beer float in a frosty mug.

"Y'all gotta upsell them floats," Mr. Bogle, the franchisee, told me about twenty times on my first day. "Talk up the frosty mugs."

Mr. Bogle had lured the day manager away from Hardee's. Eldon Otto resembled Tim Conway with no lower front teeth. He taught me how to take orders from the drive-thru speaker.

"Simportant t'speak clee-lee into da mackaphome," Eldon said, leaning over and resting his lips on the microphone. "Hi-may-he'p-ya?"

"What ya said?" a voice garbled back.

Frieda and Minky cooked in the back. Lisa and Misty cashiered up front. Becky and I swept the dining room and wiped down the tables and windows, filling in on the registers when Lisa or Misty needed a smoke break.

Lisa was borderline in almost every respect: borderline obese, borderline blonde, borderline personality. Even her haircut had a split personality: short, stiff, and styled in front, long and lax in back.

"Wow, I love your hair!" I gushed after introducing myself.

"What's that supposed to mean?" she said, her right hand curling toward her back pocket as though reaching for a knife.

"I mean, it seems like all you have to do is curl the front and you're out the door!" I said.

I was terrified but rightly guessed that if I just kept smiling like a dumb puppy, I'd be okay.

"Thanks," Lisa said, whipping out not a knife but a comb, which she proceeded to drag through the capelet of hair hanging between her shoulder blades.

"Let me know if there's anything I can do for you," I said, practically groveling on the greasy floor mat before her feet.

Misty had a spiky orange crew cut with a rainbow of rat-tails hanging from the base of her skull. From each rat-tail dangled many feathered and beaded roach clips. Misty drove to work in a primer-covered Chevy Nova with a muffler so damaged you could hear her coming five minutes away, which was just enough time for me to get busy or hide. Misty was even scarier than Lisa. Cystic acne oozed the venom that ran deep in her veins. She always came in loud and itching to fight.

"Nobody working drive-thru today but ME," she'd say, grabbing the microphone, licking it like Ozzy Osbourne, and bursting into song. "Stroke me, stroke me, STROKE!"

She whipped the male drive-thru customers into a frenzy with her suggestive, menu-based plays on words.

"Baby, you sound like you already got a Big'Un. *I'm* the one that needs the Big'Un. I got two Big'Uns up top and a Littl'Un down below."

If the fellow pulled around to the window and met her fancy, Misty would yank off her apron and say, "Goin' on break." Everyone else in the restaurant would peer through the windows to the parking lot, watching her Chevy Nova rock back and forth while Ratt's "Round and Round" blared from its radio. Rat-tails and roach clips smacked against the windows.

When it was slow, which it usually was, I treated the ingredients on hand as my own personal boredom salad bar. I would go back in the kitchen and make coffee floats and hot fudge dogs to please Minky and Frieda's adventurous palates.

"Mmmm, that's good!" they'd say, licking at mugs of vanilla soft-serve topped with ketchup and pickle relish.

My survival strategy was simple: LAY LOW. I tried to stay out of everyone's way as best I could. It wasn't easy. The employees of Bunz led messy, tar-pit lives. One touch and you were stuck, but good. I'd offer someone a friendly neck massage, and the next thing you know I was draining an abscess or using a toothpick to dig out an ingrown toenail.

Misty and Lisa fought constantly over Roby, the assistant night manager. Roby was six foot seven and potbellied, with a high-pitched voice. An unlikely heartthrob, but he was the only male under fifty at Bunz. Even I carried a periodic torch for him, discreetly, of course.

Lisa and Roby shared a trailer over in the woods behind the Primate Center research facilities. That they lived together was little deterrent to Misty's overtures. Roby seemed irresistibly drawn to her boyish frame.

In the battle for Roby's affections, Misty and Lisa constantly plotted to get the other fired. A series of events snowballed toward the end of July. Lisa swiped a twenty from Misty's till so she'd come up short at cash-out.

Then Misty stole a twenty-five gallon drum of Chili-Quik from the storage shed and planted it in the back of Lisa's pickup truck. Lisa retaliated by letting Roby finger-fuck her in the break room in front of several witnesses. The next day, Misty came to work in a t-shirt covered with decorative holes.

Roby was sitting out in a booth, interviewing a potential hot dog assembler. Misty strutted over and began to dance luridly to "Midnight at the Oasis," which was playing over the Muzak system. Her nipples poked in and out of the holes in her shirt like a Whack-a-Mole game. Roby was transfixed.

"I'll kill her!" Lisa screamed, lunging at Misty.

Becky, Eldon, and I held her back. Frieda and Minky watched through the service window, enjoying my latest gourmet creation: chili cheese fries.

Slowly, things died down and got under control.

"Am I the new hot dogger or what?" squeaked a small voice.

Darlene was very petite and demure and didn't seem like a threat to the power balance. Lisa took Darlene under her wing, giving her a perm and keeping her press-on nails safe for her while she dressed the hot dogs. The only thing odd about her was that she told some of us her name was Darlene and others that her name was Sapphire. Seemed harmless enough.

One night, three busloads of Baptists pulled into the parking lot, probably because there was no room for them over at McDonald's. We hardly ever got busy and rarely had repeat customers unless it was one of Misty's drive-thru regulars. The lines were going out the door, and everybody wanted a damn root beer float in a frosty mug. My knuckles ached from pulling the knob on the soft-serve machine.

Fifteen minutes into the rush, Lisa snapped.

"Keep your goddamn hands off my titties!" she said to a small bald customer as he held out money to pay for his order.

"What?" the man said, terrified.

"His hand just pinched my tit!" Lisa yelled at Roby. "You gonna be a man for once and do something about it?"

"Lisa," Roby said, trying to smile and frown simultaneously at the customer. "You're having a nicotine fit. Go on break. Darlene, you think you can work this register?"

"Fucker did so pinch my titty," Lisa grumbled, but the promise of tobacco and a chair disarmed her.

Darlene slid in behind the register, leaving her station open.

"You do the hot-dogging while Lisa's gone," Roby said to me.

Those Baptists ate every last drop of chili in the place.

At closing time, Roby tallied up Lisa's receipts.

"You're ten bucks short, babe," he said to her.

"Darlene honey," Lisa said tightly, "I came up exactly ten bucks short. Any way you might've pocketed it?"

Darlene bent over the hot dog station like someone had punched her in the gut. She pointed a shaky finger at Misty and whispered, "She made me give it to her."

"What?" Misty barked. "Ain't that funny, Sapphire? And here you were just telling me that ten dollars was exactly what you need to get your acrylic fills done over by Barla."

Barla was the manicurist at A Touch of Hair, a beauty salon over in Mandeville. Darlene/Sapphire had grown tired of dealing with the hassle of press-on nails. Last week, Lisa had taken her to Barla for permanent artificial nails.

The shit hit the fan.

Lisa had Misty up against a wall, throttling her. Misty's eyes bulged out and a blue color crept across her lips, but she was still yelling "fuckymuthuhfuckymuthuhfuckymuthuh," so I knew she wasn't dying.

"Stop it, just stop!" Roby pleaded.

Darlene/Sapphire flung herself into his arms for protection. Minky and Frieda, still mad that Roby had reported them to Mr. Bogle for stealing toilet paper, snuck outside and called the police on the pay phone. Other than giving them the dime for the call, I stayed out of the fracas. I was scared my glasses might get broken and I'd have to start the new school year half-blind. I was pretty sure Dad wouldn't buy me a new pair if I'd been in a public brawl.

Roby fired Misty that night but Mr. Bogle rehired her the next day. Misty's brother worked at Exxtacey Auto Detail and did lots of heavily discounted work on Mr. Bogle's Buick Regal.

Everyone still socialized together after work. I found this willingness to let bygones be bygones inspiring. You could leave deep claw marks on someone's face in the afternoon and five or six hours later, be toasting each other with 48-ounce High Octanes at Daiquiris 'n' Cremes.

One Saturday night, everybody from Bunz decided to go out to the Neon Cowboy, the bar down at the Holiday Inn. The Jimmy Buffett cover band, Cheeseburger in Garydise, was playing. I was too young to pass for eighteen, but Becky went.

I woke up at 3 AM.

Shunk shunk shunk.

Someone was trying to bust into my room through the door to the porch! Or worse: the Devil had come back.

"Sarah, it's me," Becky said through a crack in the door. "Let me in."

I got up and unlocked the hook-and-eye.

"Mom just told me to sleep out in the van like the whore I was!" Becky said.

There were no keys for the front door, so whenever we got home we had to bang and pound until someone woke up and opened it. This time even the screen door was locked. Becky had scratched on it until my mom got up, apparently just to tell her she wasn't going to let her in.

Becky was pretty worked up.

"Oh my god, Sarah, you won't believe what just happened," Becky said, collapsing onto my bed.

I opened the tube of Toblerone chocolate and broke us off a couple of pieces. Becky had brought it back from her class trip to Europe. Mom had taken out a loan to send her.

That night at the Neon Cowboy, Darlene/Sapphire started flirting real hard with Roby. I could just see her, flicking her new nails through his long stringy hair, scratching them along his neck, suggestively circling that fleshy pink mole dangling from his earlobe.

"So Lisa throws a whole vodka collins, cherry and all, right in Darlene's face," Becky said. "Darlene runs out. Twenty minutes later, she comes back and pulls *a gun out of her purse*! She's all, "I'm gonna kill you, Lisa!" So Lisa and I run to the bathroom and Darlene runs out in the parking lot and Roby chases Darlene down and wrestles the gun away!"

"Then what happened?" I asked, feeling cheated out of a terrific evening.

"Oh, nothing much. Roby fired her," Becky yawned, "then I went over to the Waffle House with Roby and Lisa and watched them suck face for two hours. They're more in love than ever."

✵ ✵ ✵

"How many kids ya got here?" Mr. Brainerd said, watching us traipse on to the front porch and through the front door of our new rental house on Jahncke Street.

"Just four, come fall when my oldest goes off to college," Mom said. "But don't worry, my children are all good students. Becky's going to LSU on a full scholarship."

Brainerd was going to be our new landlord. His nose looked like an upside-down three-leaf clover speckled with pores the size of pearl barley. A fringe of white hair shot off the back of his head like a ramp.

"Yvette Marie down at the post office told me you were running a day school back at your other place," Brainerd said.

His voice and puppet-like profile reminded me of the *Twilight Zone* episode where the ventriloquist dummy comes to life and chases Cliff Robertson through dark alleys, saying, "You're not gonna put me back in that box, are yaaaaaaaa?"

"No, it wasn't a day school," Mom said nervously. "Ha ha ha. I just babysat every once in a while. I'm gonna have to set Yvette Marie straight."

"I hoid you were looking at the Moiphy place down the road," Brainerd drawled, his body remaining still while his head swiveled slowly toward my mother. "Moiphy wouldn't let you have no day school."

He blinked. His eyes clicked dryly, like the animatronic triceratops at the Dinamation exhibit over in New Orleans. The unfortunate image of him saying "spoim" entered my mind.

"No, I'm not interested in the Murphy place," Mom said.

I could always find something exotic about a new house.

"Look!" I called from the dining room. "There's a little shelf built into the wall for the telephone."

"Uh-huh, the convoy-sation niche," Brainerd said, ruining it with his nasty dummy voice.

From the outside, the new house was a cutesy cottage painted periwinkle blue. Indoors, it smelled like the drop-off bin over at Goodwill that contained the raw, unsanitized donations you could get for 15 cents a bag if you were willing to risk lice, scabies, herpes, and seed warts. There were only two bedrooms, both with dingy green linoleum floors in the style of Abandoned Sanitarium.

"At least this place'll be warmer in the winter," I told Becky.

"I'll be up at LSU," she shrugged, looking around noncommittally.

Mom stood on the heater grate in the living room.

"Gas heat?" she asked Brainerd.

"It ain't hooked up right now," he said, "but I'll take care of that before you move in."

I could tell by the look on Mom's face that the gas would never be hooked up. Even if it were, she would find some reason not to run it.

We can't afford it.

Carbon monoxide's as odorless as it is deadly.

I smell gas. Do you smell gas? I don't care what anyone says, I smell gas.

I pictured Becky ensconced in her toasty dorm room and shivered.

Losing my fabulous undertaker's office wasn't a thrilling prospect. Still, I had hopes that downsizing would help me streamline this family's lifestyle. I was looking forward to stripping us all down to the bare essentials. Perhaps I would get my hair cut into a sleek, waifish, wash-n-wear cap to match our new ascetic aesthetic.

For as long as I could remember, I'd been forced to live in a cluttered, messy house. Our furniture was a combination of late-sixties pieces upholstered in crushed velvet or scratchy Herculon that dated back to the beginning of my parents' marriage, and "antiques" that Mom bought from the junkyard and refinished herself. Hardly great bones to strip down to, but the problem wasn't the furniture, it was what lay beneath it: a tightly packed amalgam of broken toys, stained clothing, safety pins, scrap paper, pen caps, and crumpled wads of toilet paper known by the charming colloquialism: "snot rags." It was impossible to fight this tide of flotsam and jetsam. There was no point in not chucking in my own odds and ends.

Under and between the bunk beds and the wall in Hannah and Debbie's room was the most notorious and overflowing of these cramspaces: the Stoney End.

"Girls!" I said, throwing my arms around Hannah and Debbie's necks. "What say we leave the Stoney End behind, eh? For good this time."

The Stoney End overflowed with crusty blouses, overdue library books, moldy tableware, holey socks, and a giant stuffed tiger won at the ring-toss booth at Busch Gardens. It took its name from a completely unrelated song off Barbra Streisand's *Greatest Hits Volume 2* album. "Stoney End" had been a family anthem ever since I discovered it on one of my parents' eight-track tapes. In "Stoney End," Barbra sang in a jazzy southern accent about being "raised on the Good Book Jesus," exhorting her "mama" to "cradle" her.

Hearing Barbra sing like this didn't raise any red flags. I didn't know Barbra Streisand was Jewish. Louisiana was 80 percent Catholic, a clear majority. We didn't need to split hairs over who killed Christ. The important thing was that he was dead and then alive again and now we could pray to him, and no, we weren't cannibals, just more magical than y'all because we could turn bread into a man's flesh for real.

Within a week of moving into the Brainerd house, we were back to being ankle deep in detritus. Becky and I were sharing the front bedroom, so it hadn't gotten too bad. The back bedroom was a misty jungle of crisscrossed clotheslines where Mom did her ironing, which she'd started taking in for a typically lowball figure of 25 cents a shirt. Word got around and business was booming. That price was a good 75 percent cheaper than the cheapest cleaners in town. Somewhere under the canopy of other people's hanging laundry were three twin beds, adrift in a sea of instep-torturing Fisher-Price toys. There Mom, Hannah, Debbie, and Michael slept in a first-come, first-served, musical-chairs style.

"Look what I scored at Goodwill," Mom came in the front door, brandishing a red sweatshirt with snowflakes falling on the silhouette of a deer with large antlers. Over the left breast it said *Crested Butte, Colo.*, in green script.

"Great, more crap!" I half-complained, eyeballing her latest acquisition.

Its cuffs and hem had a subtle vintage patina that could, in the right light, connote Old Money.

"Tuesday's Fifty-Cent Day," Mom said excitedly. "I tried to get them to sell me that poster I like so much but they said it wasn't for sale. Can you believe they wouldn't sell it? I offered them five bucks for it, too."

"What poster?" I asked, pulling on the sweatshirt. Not bad.

"That one hanging over the checkout stand? The one with the lady with her finger against her lips that says, 'My Fashion Secret? *Goodwill.*'"

"Thank god," I said. "Did you get the arm?"

"Yeah, my buddy David saved it for me. He works in the back, sorting. I told him to be on the lookout for a stray arm and he put it aside. Only charged me a dollar."

We needed that mannequin arm for my latest Social Studies Fair project. I wasn't finished exploring the atrocities of the Third Reich. This year I'd decided to go straight to the man at the heart of it all. The only problem was, when it came to Hitler, there wasn't much *there* there: from art school reject to megalomaniacal, genocidal dictator with barely any notable stops in between. The success of the project would hinge almost entirely on presentation.

Mom covered my project's wooden backboard with a deep, blood red felt. Black patent leather letters, cut out of a pair of boots from Goodwill, spelled out ADOLF HITLER across the top. To the center of the backboard, Mom bolted the mannequin arm, raised in the "Heil" gesture. Mom was a genius. The arm was exactly the extra oomph needed to take home first place, or if worse came to worst, second.

"Hitler was trying his best to become an artist, but his lack of talent had him living on a park bench in Vienna," I said to the judging panel, reciting what I'd memorized verbatim from my written report. "Frankly, at this point in his life, Hitler was a bum."

"Very interesting, Sarah. Thank you," said the principal. She and the other judges walked on to the next project.

"Wait, I haven't even gotten to the Beer Hall Putsch of 1923," I called after them.

"That's all right. I think we've heard enough."

A group of my classmates came through the cafeteria, looking at all the projects. The new girl, Jessica Gold, paused in front of mine. She had just moved here from New Jersey. Rumor was, she was Jewish.

"Nice arm," she said.

"Thanks. I'm Sarah, student council treasurer and literary club secretary," I introduced myself, feeling an unspoken, Old Testament, pillar-of-fire God kinship with her. "You're Jewish, right?"

"Yes," she said cautiously.

"Well," I said, cutting to the chase, "do you hate Hitler?"

One side of Jessica's mouth turned upward just a teensy bit.

"I don't know," she said. "I never met him."

I was about to teach her a thing or two when the judges approached with the prize ribbons.

They walked right past me! First place went to Suzanne Fontenot's piece of crap about the Jonestown Massacre. What? Just a drab khaki backboard and a Farberware Dutch oven full of grape Kool-Aid. Suzanne had been ladling out free samples all morning. I'd seen the judges lapping it up.

"Maybe next time you should give something out," Jessica said.

"Like what?" I asked, pouting. "Zyklon B?"

She laughed really hard. I remembered the New Girl Gap.

"Wanna come over to my house after school?" I said.

The two-block walk to my house that day went like lightning. Jessica and I couldn't stop talking. She had just been joking. She knew all about Hitler, *Helter Skelter*, and the Boston Strangler and could even teach me a thing or two about her specialty, the Green River Killer. She read! She read books! I felt myself easing into her like she was a hammock.

"Nice bumper sticker," she said, pointing to our van.

"The reason that bumper's all curled is because we had to get pulled out of the snow when we were on vacation in Crested Butte, that's in Colorado," I sputtered, pointing to my sweatshirt and trying to cover for the fact that one end of the van's rear bumper was twisted like a rotini noodle. Lately it had been even rainier than usual. The parking spot on the side of the house turned to sludge and Mom had to pay Bubba Jenkins $10 to pull the van out with his truck. Technically, it wasn't even running. "We have a brand-new maroon minivan but it's in the shop right now."

"All I said was 'Nice sticker,'" Jessica said.

"Oh," I said, hoping she wouldn't notice that the van was filled to the brim with the former contents of the Stoney End.

I looked at the bumper stickers. The one my mother had put on it said, "Focus on the Family" next to a big crucifix. The other one was mine. It said "University of Budweiser." Mom kept taking it off, and I kept sticking it back on until the adhesive wore off. Now it was dangling a little, barely stuck on with some Scotch tape.

"Sarah!" Mom called from the porch. "The fridge broke again—could you bring me that big red Igloo cooler from the shed?"

"Mom, this is Jessica, from school," I said.

"Hi, Jessica!" Mom said, waving.

An amount of hair too substantial to be dismissed as stubble shone from her armpit like a hippie beacon: *Bring me your tired, flavorless, preservative-free breakfast cereals!*

"Hurry up, Sarah—the ice cream's melting."

"Why don't you wait inside for me?" I said to Jessica, not wanting her to see the tangled mess of the shed. Letting her see the inside of our house was gonna be bad enough.

I took off my sweatshirt and rolled up my sleeves. I went around the side of the house to the shed and dug through old rusty bikes and rakes and gasoline cans. I dragged the cooler out and carried it up the steps to the house. Something slid around inside, probably a brick of blue ice.

"You know, my best friend is a Jew," Mom was saying to Jessica in the kitchen. "Can you guess who it is?"

"Who?" said Jessica.

"Jesus," Mom said, like a gentle *Boo!*

"Here's the cooler," I said, setting it down next to the fridge.

"Good, get the perishables in there," Mom said. "I'm going to buy some ice at Pic-a-Sac."

I opened the fridge. On one shelf was a covered Pyrex dish of leftover chicken stroganoff. An authentic old family recipe passed down from Grandma Pat, chicken stroganoff contained condensed cream of mushroom soup, sour cream, a can of mushroom stems and pieces, and a whole jar of pimentos.

Three days later, it looked like sad barf. I closed the fridge door fast.

"Say, let's get the cooler ready," I said, lifting its lid.

Jessica and I both jumped and screamed.

On the stark white bottom of the cooler lay Sylvester's wide-eyed corpse, twisted like a desiccated rag, her paws clawing skyward. I hadn't seen her around much lately, but Sylvester was strictly an outdoor cat. When I didn't hear moaning underneath my bedroom for a few days, I assumed she had found a new location for her sex romps. She must have gotten into the shed and climbed into the cooler, which Mom usually kept open so mold wouldn't grow inside it. She must've knocked the lid closed, trapping herself inside.

"I think I want to call my mom to come pick me up," Jessica said.

Twenty minutes later Jessica's mom pulled up in a teal minivan. As I waved good-bye, Jessica avoided making eye contact. When I went back into the house, Hannah was sitting on the foot of my bed, cutting the neck out of my Crested Butte sweatshirt.

"What the hell are you doing?!" I said to her.

"I was just making this into a Flashdance," she mewled.

I shoved Hannah down on the ground. She plucked a can of Suave Extra Body mousse from a pile on the floor and threw it at my head. I ducked and the can hit the window, cracking it.

Normally you could just feint toward Hannah and she'd take off lickety-split. She was such a spazz, she'd inevitably run into a wall or piece of furniture, thereby taking care of your revenge for you. It was a terrific way to avoid being blamed for her injuries.

"I didn't do anything. Hannah bashed into the wall herself. And besides, she started it!"

These days, I fought more with Hannah than with Debbie. Perhaps as an adaptive response to being the family punching bag, Debbie's fingernails had evolved into long talons. Her survival strategy was to latch into your flesh and hold on. My forearms were freckled with little white half-moons, souvenir scars from bouts with Debbie.

When I saw those cracks in the window, I saw a red so dark it was black. I pulled a tennis racket out of thin air and brought it down on Hannah. She curled up into a ball, bleating her piteous little bratty cries. I hit her on the legs and behind, repeatedly. Then I hit her some more. The dark molecules before my eyes began to scatter, floating off like amoebas. Slowly, slowly the tennis racket rose and fell less frequently. From the heap on the floor in front of me emerged a helpless little girl.

"I'm sorry," I said. "I'm so sorry."

I dropped the tennis racket and walked out of the house, my spine stiff as a steel rod. I walked down to Bogue Falaya Park and climbed a fig tree. Its leaves were big as a man's hands, or bigger. The branches curled around my body. I felt brittle and hollow, like a locust shell. I clung to the bark of the fig tree, hoping to draw something out of it. I sat there for two hours, until it started to get dark. I climbed down and went back home. My cheeks were raw from penitent tears.

Mom met me at the door.

"If Brainerd sees the cracks in that window, you're paying for it."

"But Hannah's the one who threw the mousse can," I said, calmly, as I would now forever be.

"Well, I don't know about that," Mom said. "You're a real bully."

Hannah peeked out from around Mom's hip, whimpering. Mom stroked her head slowly, sensually, like a Bond villain with his prized pet kitten.

Fresh tears, tears for the lack of justice in the world, flooded my face. My blissful calm was officially shattered.

"No fair!" I cried, stalling while I mentally rehearsed my defense:

I took extra care to whack her only on the thighs, the fleshiest part of the body, no vital organs. Hannah in turn showed me no such consideration. If I hadn't thought to duck and the can of Suave mousse had hit me in the head, I could've ended up in the hospital with a concussion, or worse: brain damage, at the very least a paraplegic. Bills galore. Probably have to get one of those fancy electric wheelchairs with a snack tray attachment, and that means snack expenses.

"Crested Butte," I sobbed, by now incoherent. "Chopped up."

"Ugh," Mom said, turning away in disgust. "You're such a screaming Mimi."

She drew Hannah close and they walked out of the room as though they were one. I collected my wits. I would move into the fig tree. First, though, I would avail myself one last time of indoor plumbing.

I went into the bathroom and sat on the toilet. There was a spot of blood on my underwear.

"Mom!" I yelled. "I got my period!"

A hand mirror lay on the floor by the toilet, in case anyone needed to check up on anything down there, as though a vagina were some shifty creature you needed to keep an eye on. I angled the mirror between my legs. Yep!

Mom sailed into the bathroom.

"I could tell by the way you were beating on your poor sister that you were getting ready to start," she said. "We're all out of granny pads, so just wad up some toilet paper and stick it in your underwear."

Creamin' in My Jeans

"During ovulation, vaginal mucus is thinner and more watery," said the narrator, a professional announcer with a deep smoker's voice. "This enables the semen to travel up the vagina, through the cervix, and into the uterus, where the egg is waiting to be fertilized."

My junior class at Prompt Succor was in the chapel watching a short film for our weekly class, Christian Sexuality. The visual sequence accompanying the voice-over was a thumb and forefinger pinching and stretching a stringy, viscous substance. I wondered if it was the smoky-voiced man's fingers. I didn't see any nicotine stains. Nice nails. The mucus string popped with a *sproing!* sound effect.

"Bleccccchhhhh!" gagged Desiree Pettitbon, the class illiterate. "That looks like scrambled eggs. I'm gonna blow chunks."

"Yeah, jeez yawl, this is gawbitch," Marzy Aubert chimed in.

Back in ninth grade, Marzy had gone with a guy who told everyone her vagina "tasted like a rose." Putting aside the question of whether that was the desired flavor for a vagina, I'd since assumed she was comfortable with her body in a way I could never be.

"Girls! Ssh!" said our religion teacher, Mrs. Erny.

"The rhythm method is the only acceptable form of birth control for Catholic couples," Sexy Smoker continued. "With non-Church-approved forms of birth control, things often go disastrously wrong."

Dark, spooky music blared from the projector's speaker. An x-ray of a fetus flashed on the screen. It had what looked like a tiny cinnamon roll growing out of its spine.

"What. Is. That." Joy Fink said, momentarily distracted from taping up the hem of her uniform skirt to forbidden heights.

"That's an I.U.D.," said Ms. Legendre, the other religion teacher, sounding bored.

"This is why you need to wait 'til marriage, girls," Mrs. Erny said, flipping on the lights. She closed her eyes and put a hand on the back of her blue-black hair, like she was feeling it for fever. "I had eight kids with my Albert and I still love to climb up on top, straddle him! Sex is a beautiful thing, girls. But only within the sanctity of marriage."

"She's right," Ms. Legendre said. "When I was in college, I hardly ever went to church. Finally, I go and I'm like, 'God, I need a sign that I'm supposed to be here' and right then this really cute guy, he had like, long blond hair—"

"Ewwwwwwwww!" we all groaned.

"It was the seventies!" she protested. "This cute guy comes up to me and says, 'God told me we should travel around and preach together.' So we get into his van—"

"Ewwwwwwwwww!" we groaned again.

"It was a NICE van," she continued. "Anyway, we drove all over the country, it was a really mellow scene, hanging out on the beach in California and praying, until one day he goes, 'God told me we should have sex.'"

"Did you do it?" asked Val Arthur. That morning before school I'd seen her having sex with Greg Leonard in the backseat of his car.

"No!" said Ms. Legendre. "I knew when he said God said that we should have sex, he was crazy."

"That's just like me and Greg," Val said. "We've decided to wait until marriage."

The dismissal bell rang.

"Trust me, it's well worth the wait, girls. You're gonna love sex," Mrs. Erny called. "You can't practice the rhythm method 'til marriage, but you can learn it. Don't forget to pick up your ovulation charts on your way out."

* * *

Mom came in the front door and jerked the needle off the phonograph. Hannah and I were listening to an old Stephen Foster 78, "Way Down Upon the Swanee River." It was either that or the *Star Wars* soundtrack. They were our only two records not annoyingly scratched. Mom turned off the lamp and peeked through the living room curtains.

"Ssssshhhh," she said, sitting down in the dark. "I just got flashed over in Delchamps' parking lot."

She'd gone over there to pick up some Breyer's Butter Pecan ice cream. As usual, the van overheated.

"I let it cool down while I went in to get the ice cream. A nice black fella from BJ's Pizza got a bucket of water and filled up the radiator for me. I offered him two bucks but he wouldn't take it," she said. "So then this Cajun guy pipes up, he was sitting in his pickup right next to us, and he goes, "I find it best to use half and half, antifreeze and water," and I go, "Is that so?" and the black fella says, "I gotta get back to work," and he leaves. Then I get into the van and start to drive away, but the Cajun guy yells, "Hey, c'mere!" so I circle back around and pull up next to him. He's got the door open and he was leaning over the wheel, I thought he was taking off his shoes."

I was entranced.

"So then he goes, 'Looka dis!' all Cajun, you know. And he's pointing down with one long finger down at *another* long finger, if you know what I mean," Mom paused dramatically. "Ha ha ha . . . well, it looked just like your father's."

"Duh, Mom," I said. "Dad's is like, the only one you've ever seen."

"He had a big one, your father," she nodded. "So anyway, then the Cajun, he peels out in his truck, and I start chasing him and he's going down all these dirty back roads behind Claiborne Hill, and suddenly I think, 'That ice cream's probably melting,' so I turned around and came home."

This sort of fast-lane perversity was always falling right into Mom's lap:

> *Back when I was fifteen, I went to the Kansas City Zoo with this French exchange student and her cute brother, and a total stranger came up and pinched my boob.*

Or:

> *Once when I was working downtown, right when I started dating your father, some man wearing a suit just reached out and grabbed my frontie, right there on the street, then just went along his merry way, whistling.*

"What'd you *do*?" I asked whenever she brought up her fondling history.

"Oh, I cried of shame," she would say. "Then, I got mad."

I could pass on direct, hand-on-boob-or-frontie contact, but I wouldn't mind getting a flash or two. All by itself, flashing seemed harmless. Sort of like, "Excuse me, miss. Here it is. Any interest? No? Oh well, then. I'll just be on my way. Nice doing business with you—take care!"

Strange men occasionally flitted about the edges of Prompt Succor's campus, like moths with erections. Every once in a while, some guy would jump out at a girl or group of girls and shake his dingly-dangly for all to see. Almost all. All but me. I always seemed to be on the wrong side of a tree or a building. Jean Fussell must've been flashed on seventeen separate occasions, and she had a glass eye. Not only that, her good eye was a walleye! What a waste.

Lately, a guy had been lurking behind the Grotto, a stony, cave-like structure that housed a statue of the school's patroness, the Blessed Virgin Mary, Our Lady of Prompt Succor. A recent paint job had left her robes a gorgeous azure blue, her eyes veined with bloodshot capillaries, and her fingertips buffed and polished into a French manicure.

Grotto Guy was most likely the one who'd flashed Shannon Hicks. He ran up to her, opened his jacket, and whipped it out.

"Ha *ha*!" she claimed he'd said.

Shannon shrieked and tore across campus into the locker room, looking over her shoulder like Tippi Hedren in *The Birds*. Almost immediately, she was casual about it.

"Y'all," she said, flipping her hair, "he wasn't even like, hard."

I began loitering around the Grotto. To avoid looking suspicious—or worse, desperate—I knelt, pretending to pray. Sometimes I really did pray.

Lord, please send the flasher to me. You can bet he'll get hard with me, Lord.

Weeks passed with no sign of Grotto Guy. Then, one day at lunch, I spied him a block away, furtively weaving in and out of parked cars. He stopped to blow his nose, check his reflection in a car window, and smooth his hair.

I perched on the rock wall surrounding the Grotto, trying to see him without looking. A walleye would've come in handy right about now. God, he was taking a long time. I spread my legs apart and flipped my hair back, wishing both were longer. I cupped a hand around my ear, straining to hear "Looka dis" or "ha *ha*," or something more conversational, like "Whaddaya think, babe?"

He was taking forever.

Finally, I looked up. Grotto Guy was gone. I later heard he had turned himself in. It made me feel more insecure about my looks than ever.

✵ ✵ ✵

Four of the richest girls in my junior class at Prompt Succor were having a party the Saturday before Thanksgiving.

"Join Us for an Autumn Celebration" said the engraved invitations. I had no one to go with.

Besides the kiss from Trey Blanchard and the dance with Mickey Foote, the only other dates I'd had were with John Richardson, a scholar-athlete two years older than I. We went out a few times over the summer, before he graduated and left for military school up north. Knowing we had such a short time together lent our relationship an exciting fatalism.

Within a matter of two dates, we had fast-tracked to some pretty advanced dry humping in his Chevette over at the I–12 rest area. Advanced enough to cause us to pledge to become sort of engaged.

Our long-distance pre-engagement was running low on juice, on my end, anyway. John's latest letter had petered off into a moody litany of boot-camp flashbacks:

> *Dearest Sarah,*
>
> *Today I saw myself in the mirror. I lifted the pills to my mouth. Less life, more death . . . Only the thought of seeing you at Thanksgiving keeps me alive.*

Along with the mash note, John included a recent Polaroid of himself: unsmiling, his head shaved, holding up the cover to Black Flag's *Damaged* album.

I hadn't known they were going to cut off all his hair.

Even if we could recapture the magic, John wouldn't be home until the Tuesday before Thanksgiving. I needed a date for the Autumn Celebration.

"Noelle says her brother Tommy'll go to the big party with you," Mom said.

Mom babysat for Noelle's two kids, Ashley and Melanie. Noelle's younger brother, Tommy Cusimano, was Becky's age, two years older than I. I knew a little about him. He had once dated Claudette, a reasonably popular senior at Prompt Succor. She was on the preppy/drunkard circuit, her status both raised and lowered by having a mechanic for a brother. Claudette's boobs were way bigger than mine.

I liked Noelle. Once when I'd babysat for Ashley and Melanie, Noelle took me into her bedroom and showed me the *Penthouse* issue featuring Madonna's lesbo pictures.

"Isn't that silly?" Noelle said. "She's just pretending. Fairies aren't real."

Hmmmmmmm. Maybe this babysitting racket wasn't so bad, if Mom could pimp for me on special occasions.

"What kind of car does he drive?" I asked her.

"Tommy Cusimano?" Becky asked, fluffing her asymmetrical college bob. "HA! Back in sixth grade, he cried in a doghouse at Connie Lemoine's party because she told him 'No way, José' when he asked her to dance."

"It's just a date," I said. "I'll tell him up front that I'm going to marry John in twelve years, when he's done with college and law school and his mandatory five years' military service."

Mom and Noelle brokered the deal. The night of the party, Tommy picked me up in a white Ford LTD.

"What time should I have her home by, ma'am?" he asked Mom.

"Oh, I don't know, I've never given Sarah a curfew," Mom said, flustered by his good manners. "I trust you, Tommy."

He opened the passenger-side door for me, then went around and slid behind the wheel, onto the red bench seat.

"This is my mama's car. I call it the Tampon," Tommy laughed, gunning the engine. "Cops leave me alone in this car cuz they think I'm an off-duty cop."

He reached under the seat and brought out a bottle of champagne. He handed me two plastic flutes.

"Let's take care of the pre-lims," he said.

"I beg your pardon?" I said, smoothing the black taffeta skirt of last year's Sweetheart Dance dress. *Demure.*

"Pre-lims. Pre-LIMINALS. The drinks before the drinks," he said, popping the cork. He poured with one hand while he steered with the other.

The party was at one of the rich girls' houses, a sprawling Taj Mahal reproduction. The house had several wings arranged around a long, rectangular fountain. The fountain tinkled. Music tinkled; laughter tinkled. The champagne made everything tinkle.

Except me and Tommy: we sparkled. He kept my soda can full of champagne all night. I couldn't remember ever feeling so unselfconscious while laughing, open-mouthed, until my face was maroon.

The party ended at midnight. Tommy and I sped away along the country road in the Tampon. It sparkled, too. The full moon illuminated everything electric purple, like a black light.

"Aw, damn!" Tommy said, slamming on the brakes and jumping out of the car.

He ran back about ten yards. Through the back window I watched him, his dangling shirttail glowing.

"Shit shit shit shit," I heard him say.

Tommy jumped down into the shallow ditch by the side of the road, picked something up, ripped it in half, and threw both pieces with a magnificent arm. He jogged back to the car, got in, and put it into gear.

"What was that?" I asked after a minute or two.

"A bird," he said.

"You stopped just to mutilate a bird?!"

"No, didn't you see? I hit it with the front bumper."

"And then you had to stop and rip its head off?" I said, the sparkle gone. "You're sick!"

"Sarah," he said, pulling over onto the shoulder. "I had to. It was in pain. I had to put it out of its misery."

Tommy put his hands, his bird-head-ripping-off hands, on my cheeks. I couldn't help but take a deep breath through my nose to see if I could smell bird blood. He pulled me toward him.

Four hours and two and a half bases later, I was banging on our front door. We still didn't have any keys. Finally, Mom popped the latch.

"You're getting a curfew, if this is how it's gonna be," she said, looking at me with one open eye.

"I've been knocking for over two hours," I lied, floating past her.

✲ ✲ ✲

Tommy called me the next morning.

"Let me come pick you up and bring you over by my house," he said.

"*By* your house," I teased, "or *TO* your house?"

"Huh?" he said, so genuinely mystified I dropped it. "My parents're out of town."

Tommy lived on a farm, a real farm, with cows and pigs and chickens and a barn and a tractor. Tommy had one sister and seven brothers, four of whom still lived at home. We were the only ones there today.

His parents were over in Mississippi looking at an electric pecan sheller.

"They're thinking of buying it, ten thousand bucks," he said, leading me to a door at the back of the garage. "This is my brother Daniel's room. He's working on a rig for eight weeks."

Half the garage had been converted into a twenty something bachelor-pad paradise, with a satin-clad waterbed and a piranha-stocked aquarium under a glittery popcorn ceiling.

Within a minute or two, before I could fully evaluate the circumstances, we were under the satin sheets naked. I stared at the ceiling, resisting the overwhelming desire to shriek "OWWWWWW." The ceiling looked like a doily. It reminded me of my Grandma Vivian. When she died, we got all her doilies.

"Jesus," Tommy said, rolling over. "You feel like a cement wall down there."

"Oh," I said. "Sorry."

"Hey, that's okay. That's a good thing! I just thought you were—you know, you have a boyfriend and all," he said, leaning on his elbow and playing with my hair.

"Well, we never . . ." I began, embarrassed for a million reasons: I'd never had sex before. I just had sex, real sex. I was going to Hell! How embarrassing. Oh well, Tommy would be there, too, in Hell. I was probably already going to Hell, anyway. John and I had done enough that he thought we should get married to cover for it. *We've done things most married people haven't done,* he wrote in one of his letters from boot camp. What an idiot—what we did was *nothing* compared to this. And what about this guy, Tommy? Did we have to get married now? I had met him less than twenty-four hours ago. This was barely our second date. This guy lived on a *farm.* My god. My God! What had I done?

"John and I never did it, did that," I said.

"I believe you," he laughed. "Still, you *do* go to school at Our Lady's Finest Prom Suckers."

When Tommy took me home that night, I went into the bathroom. I washed my face and pressed my forehead into the mirror over the sink, feeling self-consciously regretful, like Meredith Baxter Birney.

What had I done? I peeled off my underwear, my one pair of Calvin Klein underwear. Peach, with the signature white waistband.

Why'd you wear your best underwear then?

Slut.

Both of me agreed: I wanted this to happen. That made it worse of a sin. I opened the medicine cabinet and pulled out a rosary—they were scattered throughout the house—and sat on the toilet and said it. I swore to God I would never have sex again. I tried scrubbing the blood out of my precious designer underwear. Futile.

I went into my bedroom and consulted my ovulation chart. Eight days since my last period, which I had taken great pains to chart because it also happened to be my first period. Fertile time was day 10 through day 16. That was cutting it close.

✵ ✵ ✵

Two days later was the Tuesday before Thanksgiving. John returned from military college and rushed to my side.

"Hiiiiiiii," I said, holding him at arm's length.

No indeed, that buzz cut wasn't doing him any favors. It revealed too much of a face I hardly recognized.

"Oh Sarah, I missed you so much," he said, trying to draw me closer.

"Come on in!" I said, inserting Hannah and Debbie between us. "Let me get you something to drink. Let's see . . . would you like water or . . . how about some nice water?"

"Boot camp was brutal," John was still saying an hour later, a broken record. "You could never know . . . "

"Well, I'm sure I *could* know, if I'd been there," I said, simultaneously bored and indignant.

"I've missed you so much," he said, pawing at me. "Let's go to the rest area."

"I can't," I said, patting him. "I hate to have to tell you this, but I have a date."

John looked disappointed but not wholly surprised.

"They told us in boot camp most couples won't make it," he said, standing up, squaring his shoulders, and jutting out his chin.

"Here's your book back," I said gently, handing him his copy of Hermann Hesse's *Siddhartha.*

The next night, Thanksgiving Eve, Tommy came over for dinner. Becky's boyfriend Marck was in town, visiting from Houston. Becky had met him on her senior class trip to Europe. Marck was six foot four and rail-thin, with gorgeous cheekbones. Whenever he came to visit, he let us dress him up in cowl-neck sweaters and put false eyelashes on him. He'd put on his favorite song, Frankie Goes to Hollywood's "Krisco Kisses," and teach us how to walk a runway like Grace Jones.

Marck's enthusiasm was infectious. Before dinner, we had both him and Tommy in miniskirts and pantyhose. Hannah and Debbie fought over who got to apply their eye shadow and lipstick. Even my little brother Michael got into the festive holiday spirit, wearing one of Mom's ratty old clip-on ponytails from the seventies and dancing on the dining room table to the Buzzcocks' "Orgasm Addict."

Around 9 o'clock, Tommy and Becky and Marck and I went out and sat in the back of our van. We gouged a clearing in the transplanted Stoney End junk and set up my jambox. I popped in a homemade tape of new wave and punk music, not mentioning that it had been a present from John Richardson. Tommy could be a little jealous.

"The MINUTE I'm emancipated," Marck said. "I'm changing my name to M-A-R-C-Q-U-E."

He lined up four shot glasses from my collection, expertly mixing shots of Bacardi and 7UP.

"This is how you do it, you slam it," Marck said, putting his hand over a jigger and smacking it on the floor of the van, then quickly guzzling it down. He let out a long burp and giggled. "The bubbles fizz up and you can't even taste the rum and you get soooo wasted—Oh my god, I LOVE this song!"

Marck stood up, as best a six-foot-four person can in the back of a van, gyrating and semi-pogoing his way through the two-minute synth-pop opening of Depeche Mode's "Just Can't Get Enough."

Becky and Tommy and I slammed our shots. Then, we all slammed several more.

"You were so mean to me in grade school," Becky slurred to Tommy. "I hated you!"

"Oh, come on, you were such a goob. You used to sneeze like this," Tommy said, taking a sharp intake of breath and pinching his nostrils together with a "heemp" sound, then looking furtively in both directions.

The next song on the tape began, "Only You" by Yaz. I felt a twinge of guilt over John.

Becky and Marck stretched out under the bench seats and started trying to make out. Becky told me that whenever they made out, Marck would stop to reminisce about a humiliating experience he'd had at a water park, where he'd inadvertently given himself an enema on the jet at the top of a water slide. He slid down, plunging into the pool at the bottom after his own turds.

It would be best to give them their privacy.

"Let's go somewhere else," I said to Tommy, sitting on his lap on the van's rear bumper.

"My boss is out of town," Tommy said. "We could go borrow one of his cars."

Tommy's boss lived in Beau Chêne, the ritziest subdivision in town. Tommy found the garage door opener under the cow skull in the front yard. The door glided open, revealing the boxily curvaceous forms of an Audi and a Mercedes.

"Which one you wanna take out?" he said.

"Let's take the Audi," I said. "I like green better than brown. Besides, Hitler drove a Mercedes."

"Who?" he said, swigging the rest of the Bacardi from its 1.75-liter bottle. "Come on, let's go do donuts."

We raced along the perimeter of the Madisonville gravel pit.

"Zero to sixty in eight seconds!" Tommy yelled, turning into the gravel and pulling the emergency brake.

We spun out, around and around and around. My brain sloshed inside my skull. I bit my tongue twice. I drank my own blood. Then we did it again.

"Uggghhhhhh," Tommy said at the end of our tenth or twentieth donut. "I gotta go be sick. Pee."

He got out of the car and staggered off behind a giant hill of gravel.

I sat in the Audi, luxuriating in the still quiet of its leather upholstery, its burled maple accents. A searchlight flashed across the landscape. Was there a lighthouse around here? We were close to the place where the Tchefuncte River fed into Lake Pontchartrain. Maybe they'd built a lighthouse since I'd been out here last summer, waterskiing with my sisters off the back of Dad's fishing boat.

The spotlight raked across the car again, this time closer. I peered down the road and saw that the light was attached to the top of a car, a white Ford LTD. In one bleary, Bacardi'd moment, I wondered why Tommy had gone home and gotten his mom's car.

Shit, it was the cops. I scrunched down in the seat of the Audi. The smell of leather soaked with years of a rich man's butt sweat nauseated me.

"*Browmp!*" farted the megaphone. "Get out of the car and stand beside your vehicle."

I slid down further in the seat, my bottom hitting the floor.

"Stop trying to hide. I can see you," said the amplified voice. "I said, get out of the car and stand beside your vehicle."

I cocked up my elbow and tried to negotiate the unfamiliar, foreign-designed door handle. When I finally got it open, I spilled out on the ground, squinting into the light. An official silhouette moved toward me and pulled me up.

"Who's driving?" the cop said.

"Uhhhh, my boyfriend," I said.

"Where's he at? Vandalizing the pump house?"

"What's a pump house?" I asked.

"Don't play dumb with me," he said, leading me over to a second patrol car that had just arrived on the scene.

"He just went to go pee," I said, hopelessness setting in.

"Yeah, sure, pee, right. Listen here, young lady, you stay put," he said, pushing me by the top of my head into the backseat as he told the cop up front, "I'm going to search for the so-called boyfriend."

The first cop walked off toward the hill of gravel.

After what seemed like a long time, I saw him coming back, leading a dazed-looking, handcuffed Tommy.

"This boy was asleep out next to the pit. Snot runnin' out of his nose, woulda froze to popsicles if I hadn't found him out there."

In separate cars, they took us down to the police station behind the courthouse. Tommy was placed in a holding cell out of my sight. I sat on a desk in the front office.

"I'm Officer Freret," the first policeman said, holding out a tray. "You want a cookie?"

"Yes, sir, thank you," I said, hungry enough to shove the whole platter in my mouth. I took a smallish chocolate one and looked around. Cinderblock walls, painted seafoam green.

"This where y'all interviewed Henry Lee Lucas?" I said, referring to the serial killer who'd confessed to some murders in the area. I was hoping the strategic use of "y'all" might soften him up.

"This very room," Officer Freret answered. "Now I got a question for you: whatchoo hanging around with that clown for? Frickin' icicles coming out his clown nose."

"Oh, Officer," I laughed, reaching for another cookie. "May I?"

"Go 'head, sugar," he smiled. "Whyntchoo try callin' your mama again?"

I'd been trying to get in touch with my mother so she could come get me. Someone at our house kept picking up the phone and hanging it right back up. Probably Becky, too busy trying to help Marck change. Ever since he sent her an article about homosexuality with a note that said, *This is what I am, but I still want to get married*, they'd been working real hard at it.

The other cop came back into the room.

"That fool's brother-in-law's coming to pick him up," he said. "What's going on with this one here?"

"Can't get through to this one's mama. Guess I'll drive her home," Officer Freret said, holding out the cookie platter. "Better take you a handful for the road."

When we got to my house, Officer Freret escorted me up on the porch. Mom stood on the other side of the screen door. The red and blue lights on the patrol car flashed across her floor-length flannel nightgown.

"Is this your daughter?" Officer Freret said, putting his hand on the small of my back.

"Well," Mom said, pressing her mouth into a line, "I'm ashamed to admit that it is."

I slunk past her into the house.

"How do you think this makes me look?" she said as she closed the door. "A police car pulling up in front of our house in the middle of the night? You're never seeing that Tommy Cusimano again."

The next morning was Thanksgiving. The smell of sage and turkey flesh filled the house while we watched the Macy's parade on a tiny, black-and-white Philco perched atop the two broken televisions.

"Is that no-good Tommy coming over for dinner?" Mom asked.

"No, you told me I couldn't see him anymore, remember?" I sniffed. Then: "Can I call and invite him?"

"Well, I don't like the idea of it at all," Mom huffed. "But if you do, ask him to pick up some André pink champagne on the way over."

* * *

Dating Tommy Cusimano was an experiment in living off the land. All those years of reading *Little House on the Prairie* books came to fruition. Tommy made me feel capable and chipper and game, throwing myself into various chores the way I never did when Dad made me rake the stupid yard.

Besides, none of the preppies at St. Jude, Prompt Succor's brother school, would have anything to do with me. Except for John, whom I had blindsided at the last minute by asking him to prom in my faux-Lacoste sweater, they all rebuffed my advances. Somehow they smelled poverty, sensed I was naught but a lowly hermit crab, my Mayflower wardrobe a borrowed shell. They were Old Money; I was No Money. Dating Tommy, a veritable woodsman and farmer's son, was my only option. And now that I'd had sex with him, I had to marry him.

The sun was barely up. We'd been up since four, hosing out the hunting dog cages, spraying their shit into the adjacent hog cage, where it was promptly eaten. Now we were out in the swamp beyond the farm, shooting a 12-gauge shotgun at whatever moved: bugs, leaves, twittering songbirds. In spite of being too vain to wear my glasses, I was a damn good shot. I loved guns, loved the three-day ache their kick gave my shoulder.

"You're a good shot," Tommy said, slapping me on the back all chummy and sexy. "I got an idea—let's set some traps out in the swamp!"

He went behind the barn and dragged out a pirogue, a small bayou fishing boat that looked like a squared-off, flat-bottomed canoe. He patched a hole in it, and threw in some steel-jawed traps from off the wall of the shed. We dragged it down to the edge of the marsh.

"What'll we catch?" I asked.

"Mink, maybe," he said, lashing a mallet to the side of the pirogue.

We floated out, cracking the bright green duckweed of the marsh like an ice cutter in the Arctic. Here and there, we stopped so Tommy could tack open traps onto fallen logs. I heard a splash behind me and jumped, causing water to spill in over the sides of the pirogue.

"Aaaagghhhhh!" I screamed, water pooling up around my ankles.

"Start bailing," Tommy said. "I'll row for that sandbar."

The swamp water was a brown soup, turbid with life. A zillion microorganisms penetrated my skin. I panicked. I stood up in the boat, my feet forcing it completely down into the water. It sank beneath us. I flailed around in the thick, brackish water, convinced that frogs and snakes and gators and all manner o' swamp critters were converging upon me for a feeding frenzy. I could feel the paramecia and amoebas swimming up my urethra and latching on with their gelatinous, parasitic false feet.

"Calm down!" Tommy yelled, dragging me toward the sandbar by the arm as I splashed and skittered about, screaming in a high-pitched, *girly* way.

We climbed up on the sandbar, which was really an island completely surrounded by water. Tommy gathered whatever dry kindling he could find. He built a fire with the lighter from his tackle box, the only thing he'd been able to salvage during the shipwreck. He hung up some twine and started to undress.

"Better take off your clothes, dry 'em out," he said.

I stripped off my jeans. I could barely look at him. My secret was out: I hated nature! Even I hadn't known it until just then.

Tommy made a lean-to out of some branches and filled it with dry leaves.

"How long're we staying out here?" I asked, super-casually.

"I don't know," he said, squinting out over the swamp.

He dove in, graceful as a dolphin, surfacing every once in a while. When he found the pirogue, he dragged it back to the sandbar. It was already covered in green slime.

"Clothes're still wet," I said.

We lay in the lean-to, naked. Tommy smiled down at his penis.

"Once on a camping trip all I had to masturbate with was mayonnaise," he reminisced.

"YUCK!" I said. "You know I hate mayonnaise."

"It was either that or, what do they call that stuff, mustard and pickle relish mixed together?"

"Chow-chow," I sighed.

Tommy had masturbated into the pelt of the first animal he ever killed, a rabbit. I didn't understand his compulsion to accessorize. For me, the handheld shower nozzle in the bathtub at home was always reliable. All I had to do was lie back and try to forget God was watching.

"Masturbating's vile," I sniffed.

"You're such a girl," Tommy said.

I broke off a tiny piece of pine needle and dangled it out the side of his penis hole like a cigarette.

"Look, it's Andy Capp!" I said. Then, switching to a cockney accent: "Hey Flo, I'll be down at the pub wif me blokes."

When our clothes were no longer dripping, we pulled them on. I took a deep breath and hung onto the gunwales of the boat. We paddled back to Tommy's, getting back just in time for lunch. His mom was dishing out stuffed mirlitons and something called "red gravy."

"What is this?" I asked, chomping down on a chewy lobe of meat.

"Hawts," said Mrs. Cusimano.

"Oh," I said, filtering that through my New Orleans accent decoder. "What kind of hearts?"

"They was in a bag in the deep freeze—hawts, kidneys, livuhs—probably from that steer we slaughtered a while back," she said, ladling more onto my plate. "You want some more, just help yourself, hawt."

I ate four servings of red gravy over noodles. I then ate two servings of just plain noodles and one last chunk of heart.

"Let's go back out and check the traps," Tommy said.

"I think I'll stay here and help your mom with the dishes," I said, embracing a more traditional role.

I spent the afternoon picking and canning pole beans. Tommy came back a few hours later.

"I got two nutria," he called through the kitchen window. "Dead and alive."

I came out of the house, keeping a safe distance.

"Isn't nutria that animal they brought in to eat the kudzu?" I asked.

It was okay to talk science with Tommy, as long as it was conservation-related.

"Yep, strictly a nuisance animal," Tommy said, dropping the carcass of the dead one on the ground. "Get me a coat hanger out the box in the garage, will ya?"

I got the hanger and brought it to Tommy. I gave the dead nutria a quasi-interested look. It was brown and hairy and fat like a beaver, with a rat's naked tail. The upper lip was curled back, exposing pointy, orange incisors. They looked like small, sharpened carrots.

Tommy untwisted the coat hanger and threaded it through the dead nutria's ankle bones. He hung it upside down on an old swing-set frame in the yard.

"Don't you bring that nutria into my house, Tommy!" Mrs. Cusimano playfully yelled out the kitchen window. "I ain't cooking no nutria!"

Tommy cut around its ankles with a sharp knife. He worked his fingertips between the fur and the muscle and pulled. Slowly, the nutria's coat started to peel away, inside out, like a pair of tights. Every once in a while Tommy stopped to scrape at a tenacious bit of tendon or membrane, then pull some more. Finally, after he made a few cuts around its face, the skin came off in one bloody piece.

An insistent *scritchet*-ing started up in a milk crate a few feet away. It was the live nutria, who'd had a front-row seat for the flaying of his cousin.

"I got him in there with a piece of raw bacon," Tommy said. "Lil' fucker nearly bit my thumb off when I took him out the trap."

He bent over and checked the lid on top of the crate.

"What are you going to do with him?" I asked, tasting red gravy and organ meat in the back of my throat.

"Thought we'd bring him over to your house, show your brother," Tommy said. "I used to love things like that when I was little. Maybe we can tame him, keep him for a pet."

"Really?" I asked, doubtful.

In my experience, rodents tended to become *less* tame through contact with people. We'd had pet rabbits a few years back, and they hadn't worked very hard at being pet-like.

"They shall be called Taco and Bernard," I proclaimed in the car on the ride home from the pet store. Naming pets was always the best part, the honeymoon phase.

"I'm calling them Blinky and Miss Bianca," Hannah said.

"You can't change their names like that," I said.

"I can too!"

When we got home, Debbie stuck her finger through the chicken-wire hutch and smiled.

"I'm gonna call them Marker and Crayon," she whispered.

"No!" Becky said, swooping down like a bird of prey. "The burnt umber one should be Mr. Tudball, and the ochre one should be Mrs. A'Whiggins."

I went out to their cage every night after dinner. I would hold Taco/Blinky and Crayon/Tudball close, giving them every chance to reciprocate. Invariably, my reward was a giant yellow wet spot across the front of my shirt. Within a few weeks, the rabbits grew long, sharp fangs and bit me whenever I handled them, as if to say: "Hey, let's not pretend." Cursing, I would plop them back into their hutch and rush to the carport to examine my fresh wound.

"Yeah, that's right: run, human, run," the rabbits said telepathically. "You just go back to your people-house, and mind your own damn business."

When their cute little feet sprouted fearsome talons, I decided that these particular rabbits, though store-bought, were more suited for life in the wild. Employing a technique I'd seen on *Mutual of Omaha's Wild Kingdom,* I constructed a sling to remove them from the hutch and carry them to the woods behind our house. After everything we'd been through, after all the blood, sweat, and tears, I remained optimistic. When I let them go, I expected them to nibble at the wild onion grass, sniff around, and hop back over to me for some kindly, human reassurance. I would pat them on their heads, and they would rise up on their hind legs and tell me their innermost hopes and dreams, in English. Instead, the rabbits took off running into the forest lickety-split, making no eye contact whatsoever.

Nutria were an entirely different, potentially more grateful species. Tommy and I loaded the milk crate into the trunk of the Tampon and drove over to my house. When we pulled up, Mom and Hannah and Debbie and Michael were sitting on the porch, cooling off in front of a box fan.

"I don't know what to do about this," Mom said, pointing inside.

A nest built high up in our chimney had fallen down into the hearth and imploded. Little purple baby birds, their yellow eyelids still sealed shut, were crying on the living room floor.

"I thought maybe you'd want them, Tommy, as pets or something?" Mom asked.

Tommy knelt down in front of the fireplace. He touched one of the birds on the head.

"You got a shovel?"

Mom fetched the only shovel we had, a long, narrow post-hole digger. I could tell she and Hannah and Debbie and Michael had no idea what was about to happen, but I had a pretty good idea.

Out in the yard, Tommy dug a hole. He came back inside and scooped up the pile of noisy, broken birds. Through the window, we watched him slide them into the hole and chop them up with the digger, then bury them. Hannah and Debbie cried.

"Here's a little something we caught out in the swamp," Tommy said cheerfully, coming back into the house with the live nutria.

Tommy put the crate on our coffee table and sat down in front of it with my brother on his lap. The nutria rustled in its bed of straw and hissed. It peered out of its prison with the eyes of a creature just smart enough to know how dumb it is and angry because of it.

"Isn't that something?" Tommy whispered into Michael's ear.

✵ ✵ ✵

The girls at Prompt Succor loved Tommy. He was so popular, we rarely went out on a date alone. It might have had something to do with the fact that Tommy was over eighteen. He could legally buy booze.

Tommy would pick up me and my friends in the Tampon, carpool-style, then swing by Time Saver to buy us each a bottle of Boone's Farm Strawberry Hill wine. We'd drive around drinking our wine, then go "Bowling" and make fun of the people who were merely bowling.

One Friday night, Tommy and I found ourselves out on a real date, just the two of us. We were at Pizza Inn for their Meat Lovers Special: two medium pizzas and two pitchers of beer for $10. In the booth next to ours were three bikers, two men and one woman, smoking and drinking but not eating. They must have come for the Beer Lovers Special: four pitchers for $8.

"Y'all wanna have a downin' contest?" said one of the bikers, leaning over into our booth.

"I don't know about *that*," Tommy said, his voice the perfect mixture of macho and humor.

"C'mon now, you and your lady against me and my lady," the man said, using his wrist to indicate me, then his companion.

I leaned around the side of the booth to check out the competition.

"Mmyarr," said a medium-sized mountain of frizzy hair and leather. That was the woman. The other man was dozing, dried foam in his moustache.

I turned back to Tommy, scared but confident. Tommy had taught me a neat trick whereby you open up your throat and just pour the liquor down.

"We doin' this?" the biker said. "How's fi'e dollar?"

We brought our pitchers of beer over to their booth.

"Make it ten," Tommy said.

"Apiece," I added.

The man raised his eyebrows, the leather mountain flipped her hair back, we all raised our pitchers, and the chugging commenced.

Tommy and I slammed our empty pitchers down first.

"Shit, girl!" the biker said.

They handed over twenty bucks cash *and* treated Tommy and me to a Beer Lovers Special of our own.

Later, parked in the Tampon at the curb outside my house, I blearily attempted to perform my girlfriendly duties. All that pizza and beer decided to come back for an encore.

Tommy hosed himself off in my little brother's wading pool, next to our garden statue of the Virgin Mary.

"What'll you tell your mom about your pants?" I moaned, picking sausage out of my teeth.

"I'll just say what I used to say, back when we started—you know," Tommy said. "That I spilled McDonald's chocolate shake in my lap."

When he pointed out that our winnings covered the cost of the meal and then some, I felt instantly better.

Somehow, word of the downin' contest got out around school. The girls stopped calling me Ya-Ya and instead started calling me Meat Lovers.

I resisted telling them the singular would be more appropriate.

* * *

"If I rig this with high-test fishing line and put razor tips on the arrows, we could take down an alligator," Tommy said a few weeks later, thoughtfully examining his compound crossbow.

"Why can't we just use a gun?" I said.

I preferred guns, but Tommy had a fascination with medieval weaponry. The bookcase in his bedroom was filled with slingshots and miniature catapults, handcrafted from found materials. All of them actually worked. Tommy mastered the operation of each with long hours of practice. He'd recently spent months fashioning a Wrist Rocket out of a piece of firewood and a rubber tourniquet from a first-aid kit. The sling was made out of that sentimental favorite, the old jizm-stained pelt of Tommy's first kill. He tested the efficacy of various forms of ammunition, from pennies to pecans to chunks of cement from a burned-down outbuilding's foundation.

"Aren't alligators on the endangered species list?" I asked hesitantly, not wanting to appear anti-swamp after my freak-out during our trapping excursion.

"There's a difference between preservation and conservation," Tommy said.

The following evening, Tommy and I carried the patched pirogue over our heads for a mile in the dark, trespassing through the backyards of lakefront mansions to the bayou's edge.

Kabloosh! smacked the boat into the water.

"Quiet 'til I say," Tommy whispered. No talking until we were away from the docked yachts of civilization, obscured by mossy trees.

Kneeling in the stern of the pirogue, Tommy used a push pole to snake us through the marsh.

"Light," he said, using the shorthand we had discussed earlier.

I shone a powerful flashlight out over the surface of the water. The beam reflected off a planetarium of orange gator eyes.

"Endangered my sweet stinky crystallized cumdrops," Tommy said way under his breath.

He shouldered his cherried-out bow and took aim.

Splash!

Miss.

"Damn," he said.

"Well! That's that," I said. "Guess we'll be on our way!"

"Ssshhhh . . ." he said, dashing my hopes. "Light."

For the next couple of hours, I shone, he shot, he missed, he cursed, I hoped, we waited, and then, after he toed me in the back, the whole cycle began again.

An alligator, when its hide is pierced by a razor-tipped arrow, still thinks it's going to get away. It took about thirty minutes to haul this one in, and it was only a four-footer. It alternated between thrashy "yain't-getting-me"s and languid "fine-take-me-my-life-is-over"s.

When we finally got him boatside, Tommy made a strange gesture with his arm. I couldn't for the life of me remember what that one was supposed to mean. Finally I just said, "What?"

"Get the rope!" he stage-whispered.

"What rope?" I said.

"That yellow nylon trawling rope I asked you to bring."

Oops. A clear picture flashed in my mind of the yellow nylon trawling rope hanging on the wall of the barn, undisturbed.

"Where you been with that pirogue?" Mr. Cusimano asked when we got back to the farm.

"Alligator hunting," Tommy said.

"What? You crazy, boy? They on the endangered species list. I'da called the game warden on you and whupped you too," his dad said.

"We didn't catch one," Tommy said.

Mr. Cusimano fingered the razor tip on one of the arrows.

"You make this yourself?" he said to Tommy.

"Yeah," Tommy said.

"That's some fine work," his dad said. "Probably take down a gator."

"It did," Tommy said. "We shot one, we just couldn't land it."

He looked over at me.

"How big?" Mr. Cusimano said.

"Four feet, give or take," Tommy said.

"Enough for sausage," his dad said. "Don't tell your mama; she'll be mad it got away."

✲ ✲ ✲

When Prince came through Louisiana on his *Purple Rain* tour in 1985, Tommy and I went to see him at the Superdome. I wore all purple, down to Becky's fishnet tights. Prince looked like a speck on the stage. He didn't play my favorite song, "D.M.S.R." During the encore, some guy on the upper mezzanine dumped a 32-ounce beer on my head. The cup ricochet'd off my skull and hit the girl in front of us. She tried to claw out my eyes, but then Prince started playing "Delirious" and we had to stop and dance. It was a pretty good concert. On the drive home across the Causeway, Tommy and I chugged Budweisers and threw the empty bottles at the emergency call boxes along the side of the bridge.

But that was a special occasion, the exception rather than the rule. When it came to cultural enrichment, I leaned more and more on Tommy's older brother, Patrick. I accompanied him to Maison Blanche, helping him pick out the right long, white, silk opera scarf to go with his camel hair trench coat. He waited tables at Chung Chow and would always bring me any leftover egg rolls they had lying around at the end of the night. Patrick and I saw movies like *Amadeus, The Purple Rose of Cairo*, and *That's Dancing!*, crying at exactly the same moments. With Tommy, I went to see *Police Academy 2* and *Porky's Revenge*.

One hot winter afternoon, Patrick and I were out behind the hog pen, sunbathing. Every once in a while, I got up to fetch us fresh drinks or massage some more of a custom-made baby-oil-and-iodine mixture into his skin. It bronzed and moisturized while promoting tanning.

"You're getting darker on this side," I advised. "Flip onto your back."

Patrick held onto his skimpy nylon jogging shorts, the kind that unsnap on the sides, and rolled over.

"Give me a little sip of my drink?" he said.

I put the diet root beer up to his lips.

"You need more zinc oxide," I said.

"Oh my god, did I tell you what Diamond said the other night at work? She's a prep cook and her daughter Destiny is a dishwasher, and they're both pregnant," Patrick said, looking in his mirrored aviators while he applied the white paste to his nose.

"They're both pregnant?" I gasped.

"Yeah, Diamond's like thirty-five and Destiny's around twenty," Patrick continued. "Anyway, Diamond is soooo flirtatious, right? So I go back in the kitchen to pick up an order and she says to me, 'Patrick honey, when you coming over to try out my new waterbed?' and I say, 'Diamond, you're kidding, right? You're like, eight months' pregnant!' and she says, 'Come on now, you can *feed* the baby.' I mean, can you believe she said this to me?"

"Gross!" I said, secretly wishing my mother had been wowed enough to name me after a pricey gemstone or a synonym for fate.

"Sarah!" Tommy called from way over by the barn.

"Tommy's calling you," Patrick taunted.

"I know," I groaned.

Quickly, I lay down and closed my eyes, pretending to be asleep.

"Sarah! Sarah! Sarah!" Tommy yelled, running toward us like a pesky toddler. "Crepe Myrtle's about to give birth!"

Crepe Myrtle was the Cusimanos' prize cow, named after the purple flowering trees that lined their driveway when she was born. Tommy had raised her himself on the 4-H circuit. She'd been in labor all afternoon.

Patrick snickered.

"Somebody give that cow a C-section already," he said, snapping up the sides of his jogging shorts.

"Should've taught her the rhythm method," I was about to add, but the look on Tommy's face was one of such pure, joyous anticipation, it shamed the sarcasm back down my throat.

I pulled down my shirtsleeves and pants legs and walked back to the barn with Tommy, trying to muster some enthusiasm for the miracle of life. Crepe Myrtle was leaning against the wall of the barn, heaving, and whatever the opposite of "lowing" is. Mr. Cusimano was standing about five feet away, looking strategically at the cow's vagina.

He walked over and stuck his arm in it, all the way up to his shoulder. He groped around inside her for a little while, then pulled his arm out. He began to pace back and forth, his hands clasped behind his back, like a father in a cartoon maternity ward.

"Where're the cigars?" I giggled.

"You think this is funny?" Tommy said gravely. "Her calf's breach."

"Come here," Mr. Cusimano said, waving at me with his mucus-covered arm.

I walked over, suddenly aware of being so close to this giant living creature, a self-determining animal with a separate set of emotions from my own: a wild, unpredictable, temperamental bulk of a thing with a pulse. Right this instant, its heart was pumping blood through a mesh of veins and arteries—no wait: *two* hearts pumping blood through veins and arteries, one life inside another . . . the very thought of it made me sick.

"Go ahead," Mr. Cusimano told me. "Put your hand in. You can feel the hooves."

"Do what now?" I swallowed, hard. "I mean, put it where?"

"In her . . . *in* her," he said, looking over at Tommy as if he couldn't believe his son was dating such a retard. "In her *gentles*."

"Her what?" I felt my knees giving way.

"Her gentles!" he snapped.

Sadly, one never knows at the time what constitutes a once-in-a-lifetime opportunity. When offered the chance to reach into a laboring cow's vagina, I froze, revolted by the spectacle of life renewing itself.

"Mooooo-OOOOOOOO," Crepe Myrtle bellowed, bucking up against the barn.

Exasperated, Mr. Cusimano flung his arms up in the air, sprinkling me with a little juice. He plunged his hand back into the cow and gave a mighty pull. One little hoof popped out, black as a turd.

I don't think anyone noticed me run behind the chicken coop. Pretty sure the sound of my retching was drowned out by the Cusimano family's jubilant cheers.

The Bad Seed

"Please send Sarah Thyre to Mrs. Strozzapretti's office," the school receptionist Miss Farley burbled over the P.A.

I stood up fast and nervous at the lab table, knocking over a beaker of silver nitrate onto my notebook.

"Cool ya jets, girl," said Miss Chew, the Advanced Science teacher. She flicked her lighter over a Bunsen burner and turned on the gas. "Go ahead, I'll wipe up ya mess."

I took off my safety goggles and walked out of the classroom.

Mrs. Strozzapretti, a.k.a. Da Strozz, alias Strozzy, was Our Lady of Prompt Succor's guidance counselor. I knocked lightly on her office door, which was ajar.

"Come in," she called in her stentorian voice.

Mrs. Strozzapretti was sitting behind her desk. In one of the chairs opposite was Mrs. Clapp, my English teacher. I had a bad feeling about this.

"Sit down, Sarah," Strozzy said, gesturing toward the chair next to Mrs. Clapp's. "I've called you in here because Mrs. Clapp has requested a clearance."

"Clearance" was this year's term for disciplinary intervention. It was all about triangulation and arbitration, with the occasional rectangulation and magistration.

"This morning in class . . . Oh, I just can't take it anymore," Mrs. Clapp sniffled, holding a Kleenex against her pink nose. "She's not happy until I'm in tears."

"Sarah?" Da Strozz said.

"If she's talking about today, all I said was the apostrophe should come after, not before the 's' at the end of 'Smiths' when you're saying, "We went to the *Smiths'* house for dinner.' I just thought, you know, being an English teacher, she would appreciate my bringing it to her attention. Purely a matter of principle."

"What she's not telling you," Mrs. Clapp quavered, "is that was the *fifth* time she corrected me in one class period. She's always trying to make me look bad in front of the rest of the girls."

"Sarah?" volleyed Da Strozz.

"Mrs. Clapp, I bear you no ill will," I said. "I harbor no . . . personal malice against you. . . . I am sorry you were hurt by my pointing out your mistakes."

They looked like they were waiting for more.

"I'm, uh . . . sorry?" I said.

Leaving Strozzy's office, when I closed the door behind me, I heard Mrs. Clapp break into a round of fresh sobs.

"See what I mean?" she choked.

I guessed I wouldn't be choosing Psychology I as my final senior elective. Mrs. Clapp taught that, too.

✲ ✲ ✲

"Although it may *appear* that I am running unopposed for the office of student council president," I said, standing before the entire student body, "in truth, I have a formidable opponent."

I glanced over at the principal and the faculty. All of them smiled seriously, nodding.

"And that opponent," I said, turning back to my future underlings, "is Apathy."

I closed my eyes, basking in the reverent hush.

"In conclusion, even though it may seem silly, I ask that you still fill out your ballot. Say *YES!* to me, and *NO!* to Apathy, rather than saying nothing at all, because then you'd be saying *YES* to Apathy. . . . Get it?"

The students were struck dumb by my vivid semantic coup. Our principal rushed to the microphone.

"Sarah is one hundred and ten percent right, girls. Apathy: one disease you *don't* want to catch."

As the students dribbled out of the bleachers, the principal drew me into her arms.

"Terrific speech, very brave, very relevant," she said. "You hit the nail on the head."

Together we walked out of the school gym.

"See what I mean?" she said, pointing to an exterior pea soup–colored wall.

Someone had scrawled:

The gay groper works here

"Honestly, I don't understand it," said the principal, shaking her head.

"I think they're saying Miss Gropinski's a lezzy," I said helpfully.

Miss Gropinski was Prompt Succor's volleyball coach.

I leaned in close to survey the damage, impressed and a little threatened that someone else—and a common vandal, at that—knew the colorful verb "grope" and also knew how to turn it into a noun.

"Looks like blue nail polish. A light acetone solution should take care of it, something that'll get the graffiti without lifting the paint," I said. "I'll go to the chem lab and mix up a few different concentrations, do some spot tests."

The principal smiled so big I thought her upper teeth might shoot out and scrape me into her mouth, whole.

It was morning recess and everyone was sitting around outside in their usual formations. As I walked across campus toward the chemistry lab, voices dropped in volume or trailed off. I felt as though I were leaving safe harbor and entering shark-infested waters.

I went into the lab and hoisted the big glass jug of pure acetone from the supply shelf.

"Jesus, Sarah, that speech was the biggest suck-up in history. Everybody's calling you a brownnoser."

It was my old pal Donna Grunditz, her boobs now so huge they regularly popped the middle button off her uniform blouse.

"What's the deal?" she said, scratching her neck and arm as if she were breaking out in hives. "Everybody knows she's a big ol' clam-lapping, fur taco–diving, carpet-munching bull dyke."

"It's just an excuse to get to play with chemicals," I said, stroking a pipette and reaching for an Erlenmeyer flask.

"You're like, waaaayyy too into school," Donna said, taking one last look of disgust around the lab and flouncing out.

I used cotton swabs dipped in straight, undiluted acetone to remove the nail polish.

"Sorry, Coach G.," I shrugged to Miss Gropinski as we surveyed the results.

"Appreciate the effort," she said, clapping me on the shoulder.

I left her staring at the wall of the gym, where the graffiti was now etched permanently into the paint, faint ecru on green.

School was getting tricky. Since the incident with Mrs. Clapp, I rarely corrected my teachers, no matter how egregious their errors. Believe me, I practically bit off my tongue, holding it when Miss Beaudry wrote "Porta Rica" on the chalkboard. I might need her to fill out a teacher recommendation form for my application to Yale.

I assumed the respect of my peers was secure, that I could coast on word of my wild weekends with Tommy. Upon reflection, I had to admit that the only people calling me Meat Lovers now were the desperate, nerdy girls.

Jesus Christ. I needed to rehabilitate my image.

✵ ✵ ✵

The Sunday morning movie came on a UHF station out of New Orleans at 10 o'clock. They usually played one of a handful of old black-and-white horror movies. If I hadn't drunk too much Saturday night, I would go to early Mass so I could get home in time to watch *The Crawling Eye* or my all-time favorite, *The Devil Doll.* In the lazy post-church morning light, watching a ventriloquist's dummy pull a knife out and slash someone wasn't so scary. The horror seemed sterile and distant, like a bloodless crime scene photograph.

One morning after church, I settled into the couch with a turkey TV dinner and a glass of orange juice. I'd never seen this movie before. An adorable little blonde girl, her shiny hair braided tight and her coltish legs poking out from a crinoline, was telling her mother that she knew nothing about the disappearance of a little boy at the church picnic. She shook her flaxen plaits, the picture of innocence.

As a child I'd been a blonde, my hair lightening in the summertime to wispy white corn silk. Now my hair was a light dull brown. I'd tried to get my original color back with a spray bleach product. Blonder Than You promised natural-looking results, but it turned my hair orange, and not even a summer orange. More an autumn orange: russet, like upholstery.

"And now let's return," the announcer said, "to *The Bad Seed.*"

The little blonde girl was back, skipping through a garden. A greasy man, the gardener, emerged from a clump of bushes. He tells her he knows she killed the little boy.

"You're like me. You're smart," he says nastily. "You're smart and you're mean."

Smart and mean: that sounded like an ideal combo. Trina Graziano and Elaine Campbell were smart and mean. Trina and Elaine were beautiful and horrible, enhanced by a lusty dollop of estrogen: a floater of 120-proof estrogen, lit on fire. They fashioned plunging sweetheart necklines into their uniform blouses, unbuttoning the top three buttons and tucking the edges into the tops of their bras. Their red lace hook-in-front bras. During lunch hour, they greased up their décolletés with Clinique Dramatically Different Moisturizing Lotion, slid on their Ray-Bans, and lay out to bake on the Strip, the nickname for the concrete sidewalk that bisected campus. In low voices, like gangsters, they ripped every passing student to shreds, purely for each other's enjoyment. They lived in a parallel, amoral universe. They just didn't give a fuck.

"No way, you mean to tell me these are against the law?" Trina said, incredulous, brandishing her birth control pills in Christian Sexuality.

"Bullshit," Elaine answered, equally oblivious to Catholicism's Prime Directive.

I could get away with things I normally couldn't, if I did them with Trina and Elaine. Last year we'd been assigned to portray the birth of Christ at the Christmas concert. For two hours, we were to stand silently behind the choir, posing around a newborn baby-doll Jesus in a living nativity scene.

"Let's get smashed before the concert," Trina said, "and then stay out all night."

I called their mothers on the telephone, pretending to be my mother.

"Well hello, Sheryl/Margaret!" I drawled in a Southern society matron's voice nothing like my mother's. "As you know, the girls are doing this fabulous thing for Holy Week tomorrow, and since we live so close to school, I would just love having Trina/Elaine sleep over."

Trina called my mother, not bothering to change her voice at all.

"Yeah, Sarah should really like, spend the night over here," she said. "The girls can study, or whatever. See ya."

"What'd she say?" I asked, horrified. My mother would never fall for that.

"S'fine," Trina shrugged.

Elaine stole a bottle of Old Crow from her dad's liquor cabinet.

"He'll never miss it," she said. "Right now he's on a Chivas-only kick."

Sitting in Trina's black BMW in the Circle K parking lot, we laced our cherry Icees with the whiskey and drank them as fast as we could.

"Dude," Trina said to Elaine, "put enough whiskey in and you can barely feel the brain-freeze."

Drunk, we drove to school. We clumsily pulled our costumes on over our school uniforms. As the choir took their seats, we stumbled across the stage to the manger, taking our positions in the Greatest Tableau Ever Told: Elaine was a shepherd, Trina was Joseph, and I was Mary. While the choir droned on, we cracked each other up by poking the Jesus doll in the eyes, ears, and nose with pieces of straw. After the concert, we drove to New Orleans. We spent all night dancing under black lights at a bar in the French Quarter, shooting Kamikazes and Skylab Fallouts, Prompt Succor's white blouses and signature plaid uniform skirts rolled up and tucked into our undergarments to approximate bikinis.

The next day at school, our uniforms back in place but wrinkled, our classmates congratulated us on the cloud of booze vapor wafting from our bodies. I was paranoid, waiting for some authority figure—maybe even God himself—to punish us for violating the Christ child in his manger. Half the day passed, and still no admonishment.

In the cafeteria during lunch period, I took my tray to Sister Delphine at the dishwashing counter. The sound of plastic trays sliding on top of one another felt like a steel mallet banging against my skull.

"What's the matter?" sneered Mr. O'Connor, my favorite teacher. "Head hurt?"

The thought of disappointing him hurt more than my hangover.

"Rough night?" he continued, then winked at me. "Need an aspirin?"

Trina and Elaine were magic.

✵ ✵ ✵

When word got out that the principal had hired a man to be the new Advanced Math teacher, a mucus-testing frisson shook the Prompt Succor campus. The old Advanced Math teacher had been just another boring lady. Even his name sounded exhilarating: Fred Rush. The only other male teacher was Mr. O'Connor. It would be nice to have a hunk o' fresh meat around here: male meat, too often served rare.

I showed up early for Advanced Math and introduced myself.

"Sarah Thyre, Student Council President," I said, slapping my hand into Mr. Rush's palm. "I'm available for blackboard washing."

"Nighsh to meet you," Mr. Rush said.

First looks were not so promising. The skin on his face was loose and purple. Something that looked like a pencil eraser was growing up near his temple. Standing close, I caught a whiff of mothballs and rubbing alcohol, with a top note of mint. I stepped back; no, I reeled. There was an economy-sized bottle of Listermint on the corner of his desk, the way some teachers kept a vase of artificial flowers or an inspirational plaque. Still, I kept an open mind. Surely his personality would redeem him.

The rest of the class clattered in and took their seats.

"Weekends, well—you'll find me over in Mississippi, working as a top-secret consultant at NASA," Mr. Rush told us.

He paused, allowing that tidbit to sink in while he sucked some excess saliva from his cheek folds.

"Yep, it's true. In the court of intergalactic justice, I'd be declared guilty—of being a space head! Go ahead and call me what I am: space head."

"Space head?" spat Desiree Pettitbon, illiterate as ever. "I think I'm in the wrong class. Isn't this Pre-Math II?"

"Space head?" Mr. Rush said, waving Desiree out the door. "Oh, it's a term for people like me. Some say spacists, but over at NASA we prefer the term space head."

Having Mr. Rush as a teacher was like having class outside every day! A well-timed question could easily distract him from boring ol' sine this, cosine that, and cotangent what-have-you.

"Mr. Rush, Mr. Rush! How do you go to the bathroom in zero gravity?"

"Is Han Solo gonna look as good when they thaw him out?"

"Tell us again how you were supposed to be on the Space Shuttle Challenger but your car broke down on the way."

After a couple of weeks, Mr. Rush's inconsistent answers transmogrified our curiosity into suspicion.

"But Mr. Rush, yesterday you said Neil Armstrong's favorite band was the Beatles."

Sisterhood is powerful. We were a class of teenage girls in the prime of pubescence. Our eggs were hatching and the chickens were coming home to roost.

"That dude is so not working for NASA," Elaine said one day at lunch, building a log cabin out of french fries.

"Yeah," I said. "He's such a total phony."

The Catcher in the Rye was the one book it was acceptable to let on you'd read, even more than once. Also, calling someone a phony never failed, because everyone knew that everybody else was a fucking phony.

"Somebody oughta take that dude DOWN," Trina said, eviscerating a plank of chicken-fried steak.

A plot was hatched. I made sure that my role was a vital yet minor one: lookout. I needed a piece of the action without the risk of besmirching my permanent record. Just a little something to put the kibosh on those brownnosing rumors.

The next day in Advanced Math, Trina raised her hand.

"Can I go to the can?" she asked.

"Certainly," Mr. Rush said.

"Me too, Mr. Rush. I need to go to my locker first, for some sanitary supplies," Elaine said.

"Yep!" he said quickly.

Mr. Rush was easily cowed by any mention of menstruation.

"You on the rag?" "I'm on the rag!" "She's totally on the rag," we'd gabble, waving tampons under his nose, savoring the sight of his face turning eggplant.

Instead of going to her locker, Elaine went to the office. She asked the receptionist if she could use the phone. Elaine had once attempted suicide by taking twelve Extra Strength Tylenol, so Miss Farley let her make personal calls whenever she liked.

Meanwhile, Trina went to the pay phone in the locker room and called the office.

"Miss Farley," Elaine said when Trina's call came through, "there's somebody on Line 2 for Mr. Rush. It's NASA."

Trina left the pay phone off the hook, dangling. She and Elaine hid in a stall in the restroom, probably with limbs intertwined.

Back in the classroom, Miss Farley's timid voice came over the P.A. system.

"Excuse me, Mr. Rush, but there's a very important call coming through from NASA for you," she said, sounding impressed.

Mr. Rush jumped, dropping the chalk.

"REMAIN CALM!" he shrieked. "It's probably just some sort of dire emergency they need me to handle."

He bolted out the door. I counted to thirty, then ran to fetch Trina and Elaine. We burst back into the classroom, laughing. I laughed extra hard to drown out the panic throbbing in my chest. Seamlessly, melodiously finishing each other's sentences, we told everyone what we'd done. While the other girls watched, Trina and Elaine and I took turns looking skyward, applying Cover Girl turquoise eyeliner to the innermost rims of each other's eyelids.

Ten or fifteen minutes later, Mr. Rush returned. Our heads fell over our notebooks, engrossed as never before by proofs and theorems. I felt his eyes upon us, expectant. Shit, we hadn't constructed an after-plan. Shouldn't we be peppering him with questions? We'd merely cooked up that fake phone call and called it a day.

I peeked at him over the edge of my notebook. His face rearranged itself into an expression of resignation. He reached for the Listermint. He shook his hand like it had a spider on it, ran it through his hair, and then rested his face in it. Almost imperceptibly, Mr. Rush's body collapsed, as though his bones turned suddenly to powder, pulverized by the shockwaves coming off his radar for girls' inhumanity to man.

You're smart and you're mean, I thought to myself, with zero satisfaction. My eyes itched. Damn, that turquoise eyeliner was irritating.

I had to save Mr. Rush, or at least any affection he might have for me.

"Wait—Mr. Rush—what . . . was it, they wanted?" I said. "NASA?"

"Let's just get back to today's lesson," he said bitterly.

He cleared his throat and turned toward the chalkboard.

We didn't know a good thing when we had it, and now it was too late. Mr. Rush knew that we knew that he knew that we knew he was a fake. From now on he would only talk pre-calculus. This détente could have lasted the rest of the year if Sally Comeaux had known how to keep her trap shut.

"Y'all, I don't think we're learning what we should be learning," she'd whisper in the locker room, her eyes widening in their nests of clumped mascara.

O Sally. Smart enough to feel cheated out of learning, too dumb to enjoy the free ride. Didn't she know she would get an A if she just batted those clotted Twiggy lashes and asked Mr. Rush for details about his planned journey through the center of the Crab Nebula?

Sally tattled. Mr. Rush quietly disappeared and was quickly replaced. Our new teacher waddled into class and began to write on the chalkboard. I took it all in: the bald spot, the sofa cushion–size behind, the strange metal staple in one of her earlobes. So much to work with—literally—and that was just the rear view. The glowing embers of sisterhood's unite-and-destroy instinct flickered inside me.

When I looked around the classroom, everyone else was watching the new teacher's back with surprisingly blank expressions. Did having so much to work with put this woman over the edge, out of harm's way? Yes, yes, of course, they were right: she was simply too easy a target. No pathos, all bathos.

"That's my name," the new teacher wheezed, turning to face the class. She grasped the edge of her desk for support. "Don't wear it out."

On the chalkboard, in cramped, spidery little letters: *Mrs. Cuntz.*

✵ ✵ ✵

"Don't get me wrong, Sarah," Mrs. Strozzapretti said, peering at me over her reading glasses. "You're the smartest kid in your class."

My chest swelled. Now this was the kind of guidance counseling I liked! I pressed down on my thighs to keep myself from vaulting over her desk and French-kissing her.

Da Strozz bore into me with her hard brown almond eyes. The red plastic frames of her glasses clung like a drop of snot to the tip of her nose. I held my breath, waiting for them to fall off.

"I'm not gonna lie to you—" she began.

"I wouldn't want you to," I assured her.

"I've been looking over your college questionnaire," she said, her almonds rotating down like compass needles to the open file on her desk.

With her middle finger, she pushed her glasses up onto her sharp cheekbones. I took a breath.

"I know I put Sweet Briar as my first choice," I said, "but I'm flexible on the order of preference."

Using *The Official Preppy Handbook* as my guide, I had filled out the college questionnaire with my dream schools: Sweet Briar, Sarah Lawrence, Amherst, Georgetown, Yale, and the University of Virginia.

"I'll probably miss Salutatorian by a hundredth of a point on my G.P.A.," I reminded her. "I would've made it if the principal still allowed Co-Salutatorians."

"The application for Yale alone is at least sixty bucks," Mrs. Strozzapretti murmured.

"Oh, I just threw that one in there," I said. "I figured I should apply to at least one Ivy League."

"Tut-tut-tut-tut-tut." She scanned my file, ignoring me.

I really wanted to go to Sarah Lawrence, so I could wear leotards and flowing scarves and major in modern dance. But I'd settle for the University of Virginia, whose Georgian columns and red brick buildings would surely inspire the nascent doctor in me. Or lawyer, or professor of English literature: any of the top-tier occupations from the board game Life. Anything to get me into Millionaire Acres, anything to keep me from ending up in the Poor House eating ketchup sandwiches.

"Everything seems to be in order here," Da Strozz said, almonds up. "And what makes the most sense for you is a state school."

"What about all my extracurriculars—being student council president and Spanish Club vice president and historian-at-large for the Spirit Club?" I asked, hoping my desperation sounded like perkiness. "And the 100 service hours I earned doing Clowns for Christ?"

Clowns for Christ was an after-school service program at Prompt Succor. We put on clown makeup and went over to Sunset Plantation Nursing Home to "cheer up" the old folks. From the very first time I did it, I felt cheap and phony afterward. I hadn't enjoyed cheering up my own Grandma Vivian, so why should I enjoy holding hands with strangers? It was made all the more grotesque by painting on a huge smile—literally. Over successive visits, I gradually decreased the amount of clown makeup I wore, until finally I was down to just light mascara, natural-colored lipstick, and a little spot coverage on dark circles and blemishes.

"The old people like me better this way!" I lied to the Clowns president, Bridget Landry. In truth, I hadn't noticed a difference; they were always hungry for any face-to-face interaction, be it a face covered with clown white or not.

"They're old *folks*, not old *people*," Bridget said. "And it's called *Clowns* for Christ, not . . . "

She paused, at a customary loss for words.

"Concealer for Christ?" I offered.

Bridget and I agreed to disagree. I was given an honorary discharge from the program, with Bridget signing off on my service hours.

Mrs. Strozzapretti was jotting something down on a piece of scratch paper.

"I'm sure you could get a full scholarship to a state school," she said, holding the paper out to me. "Call this number and they'll send you the forms."

"Is this because of what happened with Mr. Rush?" I asked.

"What?" she said sharply.

His was the name that we dared not speak.

"I mean, what happened with Mrs. Clapp?"

"No-ho-ho," she chuckled, pausing for a moment of fond remembrance. "It's because you really can't afford anything else. Now good luck with your application."

Shaking, I left her office and walked toward the bathroom, hoping to get there before I burst into tears. *Good-bye, sweet Sweet Briar.*

"That your college session?" a voice sneered.

It was Bitsy Marshall. Perhaps because of lingering shame over our porno past, she and I rarely spoke directly. There was always a buffer of two or three other people between us, like a preppy's layered wardrobe.

"What'd Strozzalicious say?" Bitsy asked, her voice like a raw wool sweater scratching against my naked skin.

It took all my energy to swallow my tears. I had none left to devote to lying.

"LSU," I said.

"A *state school?*" Bitsy said. "That's only all right if it's in a different state than the one you live in."

"Well," I said, pulling myself together, "it's the flagship. And Strozz is sure I'll get a full scholarship."

"Lucky for me, I'm a legacy at Newcombe," she sniffed, referring to the fashionably small, formerly all-women, recently turned co-ed division of Tulane University. "Mama went there."

As Eve was fashioned from Adam's rib, as Radcliffe was to Harvard, Newcombe was Tulane's cute, non-threatening lil' sis. I imagined a girl could meet her very own Ryan O'Neal at Newcombe, when he came over from Tulane to use their smaller, cozier library. At Newcombe, a girl could learn that love means never having to say you're sorry and still look eminently fuckable on her deathbed, riddled with cancer.

"Well, I'm also applying to Sweet Briar and Yale and U.Va.," I said, fully recovered, with lie juice to spare. "Mrs. Strozzapretti said I'm the smartest person in our class and schools'll be fighting to get me."

Bitsy's smirk burned off like morning mist. The look of awe that replaced it lasted only a second, but I caught it. A tiny, tasty morsel of victory.

✵ ✵ ✵

"Get up, it's Good Friday," Mom said, yanking down the covers. "Time to go to church."

Becky and I blinked into the sun's glare and rolled over onto our stomachs. She was home from LSU for spring break. Last night we'd gone out with Tommy. His brother Patrick had gotten him a weekend job at Chung Chow, and some of their Chinese coworkers had thrown a big house party. I had eaten tons of fried wontons and shrimp toast, drunk two bottles of Lancers white wine, dunked my head repeatedly into a garbage pail full of Jungle Juice, and then passed out between Tommy and Becky on the ride home in the Tampon.

I opened one eye, groaning in pain when the pillowcase brushed right up against the white of it.

"I want you to *dress* for church. Wear a skirt, show some respect for a change," Mom said, picking through a giant pile of clothes on the floor. "What's this?"

Mom held up my grey wool coat, pink-striped with a Nehru collar. She scratched at it and sniffed her fingers.

"Vomit? Is this vomit?" she asked. "Were you drinking last night, Sarah?"

Roll down your window, I'm gonna—I seemed to remember saying to Becky in the car. I must've thrown up all over myself instead.

"Answer me, Sarah," Mom said.

"Uhhhhhh," I said, far from my usual clever self.

"That's *my* vomit," Becky said, gouging my calf with her big toenail.

"This is *Sarah's* coat," Mom said, disbelieving.

"Yeahhhh, wellllllll . . ." Becky said. "*I* threw up on *her*."

"*You* threw up on *her?*" Mom said. "Way to go, Becky. I think you've developed a real drinking problem since you've been up at college, young lady. Probably from hanging out with those theater majors. I bet they have multiple sex partners, too."

I clenched my fist around the corner of the pillow, expecting Becky to recant, now that she was the one under attack. But she didn't.

O sweet, sweet Becky.

"Don't eat or you'll break your fast. Mass is at three," Mom said, stomping out of the room.

We got up and pawed around in the pile of clothes for something to wear.

"Thanks," I said to Becky. "You saved my ass."

"Especially considering I was the one wearing the coat—I borrowed it, remember? You threw up on *me*."

"I did?" I said, my love for her deepening.

"*I* don't have to live with her anymore," Becky said. "But you should really learn how to hold your liquor."

Mom required us to go to Mass every day during Holy Week, the days leading up to Easter Sunday. Good Friday was the day Jesus was crucified. Mass started at 3 PM because that's when he died on the cross, bowing his head and saying, "It is finished."

We walked over to St. Anne's, the church closest to our latest rental house, and sat in a pew near the back. I counted the minutes until Communion, when I would at least get a bite to eat. Just a bite, but it was better than nothing.

"Taste and see the goodness of the Lord," the congregation repeated over and over during the responsorial prayer.

My stomach grumbled.

Right before Communion was my favorite Catholic ritual. I called it Shakin' for Peace. Everyone turned to their neighbors and shook hands or hugged, whispering, "Peace be with you." The simpler, abbreviated "Peace" was an acceptable alternative. It was the only thing the least bit groovy left in the Mass, which had reverted from its short-lived hippie trappings back to somber traditionalism. Rollicking guitars were verboten.

"And now let us offer one another a sign of Christ's peee-eece," sang Father Mark tunelessly.

People turned around in their pews, leaning toward one another.

"Peace-be-with-you."

"And-also-with-you."

I shook the soft shiny hands of the elderly couple in front of us. When they reached out to my mother, she demurred:

"Oh no, really—you see, I have a little cold, just a spring bug really, but I forgot my Kleenex, so I've been wiping my nose with my hand, so I don't want to shake hands, 'cause of the germs, you know—but peace be with YOU!"

The couple recoiled, their hands going instinctively toward the middle of their chests, as if to launch a protective Sign of the Cross.

God, this was taking forever.

Communion came and went, the stingy little wafer doing nothing to assuage my sour stomach. I'd forgotten to bring a book to Mass as I usually did. *Wuthering Heights*, *Rebecca*, *Anna Karenina*, *The Art of Loving* by Erich Fromm: I got a lot of my required school reading out of the way during Father Mark's long-winded homilies. Today he was making one of his signature lame attempts to connect with the parish youth.

"In the fatalistic words of Miss Tina Turner," Father Mark intoned, "'What's love got to do with it?' Well, as the Beatles might answer: 'Tina, all you need is love.'"

I was too hung over to pursue my other favorite Mass-time endeavor: trying to make my sisters laugh out loud. Good thing, too, as it always pissed off my mother.

"Look at Dr. Camp's kids," she'd complain. "You don't see them laughing in church. Their heads are bowed in prayer."

Dr. Camp's kids have a swimming pool, I would think. *Proof that prayer works for some.*

Today, I wasn't going to stir up trouble. So far, Mom seemed to believe the vomit story.

After Communion it was time for the Veneration of the Cross, a Good Friday–only special. The giant crucifix that usually hung above the altar was taken down and paraded up the center aisle of the church by the priest and altar boys. Every once in a while Father Mark sang out, Gregorian chant–style: "Be-hold the wood of the cross on which hung the Say-vee-yor of the Wo-orld!"

"Come, let us ado-ore!" the congregation sang back.

The crucifix was laid on a table on the altar. Everyone lined up in the center aisle and shuffled by, leaning over to kiss the bloody nail piercing Jesus' feet. An altar boy used a cloth napkin to wipe off the feet between kisses.

"No way," I mouthed to Becky.

She rolled her eyes back at me in agreement.

I'd done it every year since I could walk up the aisle to the altar. I had a distinct memory of bending over, looking for the nail under multicolored smears of lipstick, then inverting my lips and smacking it as dryly as possible. This year, however, it struck me as way too gross.

Throughout the church, old ladies were whipping open their handbags and digging for their Frosty Coral or Desert Mauvestone, going for one last touchup. Men licked the hardened crust from the corners of their mouth, chewing and swallowing what they found. All ached with palpable anticipation for that annual Holiest of Smooches.

"Go on," Mom said, nudging me and Becky toward the end of the pew, "Get in line."

Becky and I looked at each other.

"We decided we're not doing it," I said.

"*What?!*" Mom whispered, as angrily as she could without drawing any attention to us.

"We think it's weird," Becky said.

"It's a sign of respect!" Mom said, beads of sweat breaking out on her upper lip, yet another thing to drip onto the bloody nail. "You have to do it!"

"It's like, the Golden Calf or something," I said. "Doesn't the second commandment say not to worship statues and graven idols?"

"You girls get in line and go kiss that crucifix NOW," Mom said, putting on her authority act.

"Sorry, Mom," Becky and I shook our heads, putting on our authority act.

"You go kiss that damn cross," Mom said, her eyes bulging out of their sockets. She looked so much like Don Knotts, the whistling theme from *The Andy Griffith Show* popped into my head.

Becky and I folded our arms.

"I can't believe you're doing this to me," Mom said, "in front of the whole church."

Stepping on our feet, she slid roughly past us out of the pew and joined the line of people waiting to kiss Jesus' feet. When she got up to the front, I craned my neck and watched her lean over. She seemed to be down there for a long time. When she bobbed up, she circled around and sat in a different pew. Away from us.

I felt like one of those people who sit through Mass but don't go up to get Communion. You just knew they'd done something terrible, something that made them feel unworthy of eating a piece of Jesus' body.

Finally, Mass was over. Becky and I walked down the steps outside, looking for Mom. I nodded politely at Father Phuc, the Vietnamese exchange priest. Initially, everyone avoided addressing him by name, assuming it was pronounced phonetically. Finding out it was pronounced "poo" didn't much change things.

"There she is!" Becky said, pointing.

Mom was pumping her fists, speed-walking away from us down the road toward our house. When we caught up to her, flanking her, she flailed her arms as though she were shaking off leeches.

"I've never been more ashamed in my life," she said, vibrating. "The devil made you do this to me."

She stalked off down the road. Following ten paces behind, Becky and I engaged in merry, animated chatter so that drivers-by wouldn't notice how much our mother hated us.

"Wait a minute," I said to Becky. "If Jesus died at 3 PM today—Friday—and he's out of his tomb by dawn on Sunday, how can they say he rose from the dead after three days? That's barely a day and a half."

"*Sarah!*" Mom said, spinning around. "Don't!"

She looked at me so sharply, her eyes might as well have sliced off my tongue. Becky put her arm in front of me, protectively, like she was driving and she'd slammed on the brakes and I was in the passenger seat. We stood like that for a while, until Mom turned away. I watched her walk the one remaining block home.

She looked old.

When Becky and I reached the front porch, I could hear our mother inside, blowing her nose.

Say Uncle

"I know your mama?" the lady said, tapping her blonde helmet of hair with a Bic pen. "I just love her. You know she's a saint, don't you?"

"Uh, yeah."

It was the middle of May, 1986. I was at Pic-a-Sac, the local grocery store that specialized in liquor and deep-fried turkeys, applying for a summer job. They had an opening for a cashier and I looooved pushing buttons. Perfect.

"I'm Myrtis. I'm head cashier," the lady said. "Are you the one who wouldn't venerate the cross on Good Friday?"

"No, ma'am," I smiled, "that certainly wasn't me."

"Oh, well then, were you the one what stayed out 'til four in the morning, making your poor mama sick with worry?" Myrtis said, tamping out her Benson and Hedges Menthol 120.

"No, ma'am," I said.

Mom had already been by this morning for milk and eggs, apparently with a side of chitchat. Pic-a-Sac was two blocks away from our house. Members of our household made several trips there throughout the course of the day, buying the makings of each meal as money trickled in.

It seemed as good a place as any to get my last summer job before heading off to college at LSU. Pic-a-Sac was close to home, and we had no car. They might have an employee discount.

"When can you start?" asked Myrtis.

"After graduation," I said. "Anytime from this coming Sunday on."

Saturday I graduated with honors, third in my class. I stood on the altar to read the second gospel in the baccalaureate mass, from the book of Ephesians:

> May the eyes of your hearts be enlightened, that you may know what is the hope that belongs to his call, what are the riches of glory in his inheritance among the holy ones, and what is the surpassing greatness of his power for us who believe, in accord with the exercise of his great might, which he worked in Christ, raising him from the dead and seating him at his right hand in the heavens, far above every principality, authority, power, and dominion, and every name that is named not only in this age but also in the one to come.

After the ceremony we stood outside the church: me, Tommy, Mom, Becky, Hannah, Debbie, Michael, and a satellite circling ten feet away: Dad.

"You sure are smart," Tommy said, fingering my gold Honors stole. "I'm so proud of you."

"My hair fell," I said, feeling the stringy, lifeless locks hanging around my ears.

Dad snapped some pictures. Milestones were always uncomfortable. My parents didn't speak to each other; we kids had to travel back and forth between them. We ended up hovering closer to Mom, half-smiling at Dad, and discussing in low voices how we wished he'd just leave because he was making us feel guilty for not talking to him more.

"What're y'all doing now?" Dad called. "Got any plans?"

Mom nudged us toward him.

"Maybe he'll take you all out to lunch at Chung Chow."

"Probably not me," Becky said, "I'm over eighteen."

* * *

The next morning I started work at Pic-a-Sac.

"Get yourself a smock out of that box in the break room," Myrtis said.

I pulled out one of the red buttoned polyester smocks with a scalloped Peter Pan collar. All the ladies at Pic-a-Sac wore one, accessorizing with brooches and large buttons that read, "I'll Cash Your Payroll Check!"

"Can I wear this instead?" I asked, tying on a mannishly tailored cotton apron that the butchers and stockboys wore.

Myrtis looked up from her compact of Corn Silk foundation. She wore her smock with nothing underneath, the top three buttons undone.

"Why you wanna look like a boy, sugar?" she asked.

"Yeah!" said Myrtis's husband, L.W., leaning over to bite her neck as he passed by.

"You rascal!" she cackled.

L.W. was the produce manager. He looked like a grayer, gaunter Johnny Cash, and wore yellow skeet-shooting glasses that matched the yellow packaging of his Price Breaker cigarettes. L.W. had the most appetizing b.o. I'd ever smelled: onions and beef and cave-ripened cheese, rich and concentrated as reduced French onion soup.

My first day on the job was a drag, like all first days on all jobs. Everyone assumes you're stupid.

"So honey, things you might eat would be what you call *gross-ry*, so you push the button that says *gross-ry*," explained Venetia, a tenured cashier with gold teeth. "Things you wouldn't eat, cleaning things like Ajax or Clairol Herbal Essence, would be *non-gross-ry*. Now let me show you how to tell *meat* from *dairy*, and ooh, *produce* is tricky. It'll take a month to tell you about produce."

"I think I've got it," I said.

A line of about ten customers had accumulated. I thought Venetia would be impressed with what a quick study I was. She left me alone for a few minutes. I whipped through seven of the customers.

Venetia came back with Myrtis and L.W.

"Says she don't need to train no more," Venetia said, pointing at me with a long, curved, orange fingernail.

"Honey," Myrtis clacked her teeth. "Training period lasts three days."

"You can't learn it all in twenty minutes," L.W. said ominously.

In spite of our rocky start, Venetia became my ally and soul mate. I loved the way she creamed orange rouge over her cheeks, from her jaw line all the way up to her eyes. At quitting time, we checked each other's gross-ries out.

"Did you happen to notice how much this turkey loaf was?" I'd say. "I'm pretty sure it's nine cents. And I *know* this 32-ounce ketchup's on sale for a nickel."

Venetia was my audience. She laughed and nodded nervously at my jokes, the way you might skirt a crazy down at Bogue Falaya Park. I didn't mind being the freak show. For me, it was a point of pride to stand out as different from the Pic-a-Sac crowd. I never put a toothpick in my mouth, no matter how good it looked.

I always suspected poverty and intellect were what barred me from being voted onto Sweetheart Court at Prompt Succor, even though Becky had made Sweetheart Court and she was smarter and looked poorer than I did. I still tried to cover both up, with varying degrees of success. Getting caught behind the counter of Pic-a-Sac dissolved the last shreds of my dream about making my debut into polite society.

Former classmates came through my line, using fake I.D.'s to buy Coors Party Balls and suitcases of Sun Country Wine Coolers, chatting gaily about graduation gifts and summer trips to Europe or the months they planned on spending at Carolinian island estates that had been in their families for generations.

"What'd you get, a Beamer?"

"Bird on you—I got a Jag!"

"The Pink Palace on Corfu totally rules!"

"So Jen says to me, 'We have a condo in Destin,' like it's a beach house on Pawleys Island! White trash . . . "

They pretended not to know me.

"Hey!" I wanted to say. "Remember me? I beat Apathy in the student council election!"

But all I said was,

"That'll be $28.47."

I began to regret not letting Mom take out a loan to send me to Europe. She'd offered. For my graduation present, I'd ordered two t-shirt dresses from a new mail-order catalog. In the world of J. Crew, fresh-faced models wearing soft cotton jersey and heirloom madras fell into each other, laughing at family barbeques on the beach. Wearing what they wore was about as close as I would be getting to Pawleys Island.

Sure I felt sorry for myself, but I felt more sorry for my old classmates. It couldn't have been easy for my fellow prom suckers, seeing me—their former student council president—grappling with dripping, bloody rump roasts and rolling 48-ounce jars of Shur-Fine mayonnaise across a UPC scanner. Poor girls were accustomed to seeing me wrestle with Apathy. If I could end up at the Pic-a-Sac, no one was safe.

Pic-a-Sac was one of the last independently owned and operated businesses in town. A few years earlier, a superstore had swept through, razing almost every local business in its path before settling out near the highway like a gluttonous, fatty beast. Why settle for Pic-a-Sac's overpriced, aphid-riddled produce and sad, dented cans of Trappey's crowder peas? Out at the mega-mart, there were gleaming vats of apricot facial scrub bubbling next to taut, sinewy hamsters and bold radial tires, sleek shotguns and crispy taco shells, all under one gigantic super-roof.

The craze for buying in bulk left, for Pic-a-Sac, a mere dribbling of customers who were either (1) too poor to own a minivan, (2) too old to negotiate 75,000 square feet of merchandise, or (3) too drunk to notice they were paying desperately high prices. Regulars.

Regulars buy the same things at the same time, week in, week out. Even the drunks were like clockwork—especially the drunks. They get a bad rap, drunks. Unreliable? They're the most reliable folks in the world.

I could set my watch by one elderly lush. She came in every Friday at 5 sharp for a box of corn flakes, a gallon of milk, three suitcases of Budweiser, and the largest bottle of baby powder we sold. She batted her blue eyes at everyone and had a kind word for all. When she fell on Aisle 2, or Aisle 7, or under the sign that said Fancy Wines and Liquors, a powder cloud poofed into the air.

"Hwy, thank yew, my dear!" she'd say to whomever helped her up off the floor.

"Now that's a real lady," Pic-a-Sac's owner Mr. Ted always said. "She'd never try to stick it to me with a slip-and-fall."

The unofficial town drunk was known as Red Man, because his skin was—get this—red as a beet. He had blond hair that verged on green, as if he bleached it, then swam daily laps in a highly chlorinated pool. One morning, soon after my training period ended, Red Man came in to the store, pacing back and forth, fretfully clutching his sweaty forehead and waving a bottle of Thunderbird Rosé.

"I know it sounds crazy," he began, not very optimistically, "but I bought this fine bottle of delicious wine last night, and when I got it home, I realized the bottle was empty! Ain't that funny? Imagine that! So, I'm just hoping I can go grab myself a replacement bottle off the shelf, one that's full, and we'll call it even."

"Wait just a minute, sir," I said.

Last night, I happened to be riding with Tommy by McDonald's, where I had personally seen Red Man out by the Dumpster, standing next to a chunky pink puddle and looking rather unwell.

I gave Red Man a parental look, the kind that said, "Now tell me what really happened."

He blinked several times at the space above my head.

"So, I'm just hoping I can go grab myself a replacement bottle off the shelf, one that's full, and we'll call it even," he said, impressively verbatim from his earlier riff.

"I'm not authorized to give you an exchange without a receipt, sir," I said.

"Norm'ly I'd provide a receipt, proof of purchase, y'see," he said. "I got respect for proper documentation."

Red Man put the bottle under his arm and pulled out his wallet, prying it open—*sssrrrrick*—with his fingernail.

"Now I just got out of the hospital, you see, 'cuz I fell three stories down and got pneumonia, y'see, and they had to put a metal plate in my head. Let me just show you. . . ."

I was afraid he was digging around in his wallet for supporting evidence, a bloody bandage or a piece of skull. I was about to give in and let Red Man go get a new bottle of Thunderbird, but I could feel Mr. Ted watching me from up in the manager's booth.

"Sorry, sir," I said, stony-faced.

"Wha, falp?" Red Man said, crestfallen. He walked out of the Pic-a-Sac, a more broken man because of me.

"Ya did good," Mr. Ted called to me from the safety of his cozy nook.

A disproportionate number of Pic-a-Sac customers seemed to have missing or mangled body parts. The first few times, I was caught off guard.

"Agghh!" I'd say when I'd state the total and reach for payment, only to find the money pinched in a metal hook or between two gnarled demi-digits, or lying flat on something that looked more like a foot than a hand.

I felt bad, startling the customers. I quickly finessed making "Agghh!" sound like the first note in a happy little song. If I had also jumped away in surprise and/or horror, I simply segued into a happy, Irish-style jig. After a couple of weeks, I recognized the amputees on sight and could brace myself for contact. By that time, an armless guy named Charlie Sprow thought the song-and-dance routine was part of my service, a personal touch. He got in my line, even though there was no waiting at Venetia's register.

"Hey, Mr. Sprow," I said, unfazed.

"Turn around for me, darlin'," Charlie said, tapping his foot. "Like you used to."

I gave a little twirl, singing along with Randy Travis on the store radio: "On the other hand, there's a golden band . . . "

Charlie let out a low whistle, his version of applause. He held down his checkbook with one shoulder and wrote in the amount with a pen he held in his mouth.

In addition to entertaining new people, I was learning a lot about food stamps. For years, in stores all around town, I'd seen signs proclaiming, WE TAKE FOOD STAMPS. I hadn't realized they were like money that you could use to buy food—any food you wanted, even name brands and luxury items.

Food Stamps 101: Separate steaks, chops, and frozen lobster tails from booze and cigarettes; pay for the former with food stamps and the latter with ancient, dirty change. Other than handling greasy coins, their moisture of unknown provenance, I had no qualms about this. We all need to enjoy ourselves. I wished Mom would swallow her pride and apply for food stamps. I could use some more steak in my life.

Food stamps came in whole-dollar denominations of $1, $5, $10, and so on. By law, I could give cash change back only on the $1 food stamps. The larger denominations got change back in the form of food stamps. Some customers, like the Hetrick family, showed real ingenuity, not to mention perseverance. They would sit out in the parking lot for hours, sending in their seven kids one at a time. The kids came through my line over and over, buying single pieces of Brach's Pick-a-Mix candies with $1 food stamps. The candy was sold by weight, and each piece cost a couple of cents. I was giving them 95 to 98 cents change per transaction, in coin. The kids brought the change out to their folks, and when they had enough, Mr. and Mrs. Hetrick came in to purchase two bottles of Mad Dog 20/20. Candy and booze: everyone was happy!

Mr. Hetrick was seven feet tall, shirtless and ornately tattooed. Mrs. Hetrick was about half his height and 300 pounds. She wore midriff-revealing t-shirts with no bra, free as a bird. Sometimes, as she purchased her usual bottle of orange wine drink, her nipples hung out the bottom of her shirt and dragged along my conveyor belt. The Hetrick kids looked like Dickensian street urchins, with blackened feet and sooty faces. My heart felt an instinctive jolt of pity for the three-year-old, until I heard him negotiating a candy swap with his older brother.

"Fine then, tittyhead, I'll give you two pissfucking Royal Nougats. Now gimme three Chuckles, cockass, and leave me 'lone, fuckle," he said, then sunk his feral teeth into his brother's thigh.

I liked customers. When there were no customers, Mr. Ted would materialize from thin air.

"Look busy," he'd say, standing two feet away with his arms at his sides.

I spent a lot of time touching the candy and magazines in their racks right near the register, patting down things that were already in their proper place. I stole roughly a pack of gum a day, chewing wads so big I accidentally bit one side of my lip. It swelled up big as a grape, creating such an obstacle I kept biting it over and over. Pretty soon the corner of my mouth looked like a hunk of raw hamburger.

"What's wrong with your lip?" Mr. Ted asked.

"I bit it chewing gum," I said.

Mr. Ted looked at me, then at the gum rack. Busted.

"No more gum," he said. "Look busy. Ya got nothin' to do up here, go back and slice meat in the deli."

I would pat gum and magazines until my palms bled rather than go back in the deli and lose a finger on the slicing machine. The rusty blade spun around off-kilter, like a warped record on a turntable.

Deli. I didn't like that word. It sounded too much like *smelly, belly*, and worst of all, *jelly*. These words had found a place on my Most Unfavorite Word List, along with *supple, fondle, nipple*, and *pleasure*. So, the people working back in the deli had a lot going against them straightaway, purely in rhyme. I prejudged them to be lurking with smelly-jelly-belly evil intent, dirty-minded and ham-handed.

I was afraid of the deli ladies. They were smokier and more disgruntled than any other faction at Pic-a-Sac, slicing angrily away at boiled hams and slathering—WHOA!—way too much mayo on things. The mayonnaise was the clincher. I still loathed it, deeply. The quantity on hand back in the deli led to inevitable thoughts of an egg-y mayo cloud enveloping my body, gurgling into my ears and nostrils, suffocating me.

Why always the huge jar, as if mayonnaise is something to be stockpiled in case of flood or famine? Why not a humble, dainty jar more befitting a so-called condiment, like the jars containing mustard or capers? It's the rare deli foodstuff unglistened with a mayo sheen or unsliced by a knife tainted with the noxious spreadable.

"Sarah, you've got to get yourself one of those four-napkin roast beefs," Venetia said, licking her fingers after indulging in an obscene pile of a sandwich. It looked like run-over entrails frosted, of course, with generous squirts from the deli's industrial mayonnaise pump. I had long suspected the pump was actually a tap, a direct conduit into St. Tammany's secret, underground, municipal mayonnaise reservoir.

The deli ladies didn't like me because I never ate their food. The fluorescent light in the glass deli case flashed like lightning, a portent. A dark film had descended upon the tub of tuna salad, and the slick yellow gel-caps of sugared ham were strewn with sad, stranded cherries. Maraschino cherries, the innately happy fruit, had to be worked over considerably to invoke melancholy.

One mid-June afternoon, a customer marched into the store and up to my cash register. She slammed down an almost-all-eaten ham 'n' cheese pull-apart. One of Pic-a-Sac deli's bestsellers, the pull-apart was a flower-shaped loaf of bread with any variety of toppings baked into its top surface.

"This was just fine and tasty, until I got to the center," the lady said, trying to contain her outrage. "Then, come to find out, THERE'S A WORM IN THE MIDDLE OF IT!"

Knowing the questionable surroundings from whence it came, I wanted to call her a fool for buying it in the first place.

"I'm terribly sorry, ma'am," I said instead, as per the dictum *customatum semper rightum est.*

I led her over to the deli, quietly, so as not to rouse the creatures within.

"Please pick out a replacement," I said, fanning my hand over the bakery rack.

The woman picked up a cinnamon sugar pull-apart.

"Hnnnn," she said, digging her thumbs into it, then tossing it aside.

A hair-netted head surfaced behind the deli counter, exhaling smoke through its blowholes.

"How 'bout this one?" I said, scooping up a jalapeño 'n' cheese pull-apart and thrusting it into the customer's arms.

"Well," the woman said, " I SUPPOSE this will HAVE to do."

She stomped out of the store, her thick ankles spilling over the tops of her navy crepe-soled shoes. I looked down, finding the remains of the wormy loaf in my hands.

Hair rose on the back of my neck. I heard a layered, burbling noise, like humpback whales calling to each other deep in the sea. Behind the deli counter, three pairs of eyes glared holes into me. The deli ladies crept closer, muttering around their dangling cigarettes.

"Whutchee sayz?"

"A worm?! Shiiii-it . . . "

"Ain't no worm in 'at."

"Djoo tell Mr. Ted?"

"Shiiiii-YIT."

"Gimme dat."

The deli ladies grabbed the pull-apart, sniffing and fingering it into pieces, which they pronounced definitively wormless. In synchronized balletic movements, they put their cigarettes out underfoot, shifted their weight, and devoured every last crumb.

Donald was a Pic-a-Sac regular, but he wasn't a drunk. He came in at 4 PM every day to buy a Coke out of the vending machine near the foot of my register.

"This is the dirtiest Coke I ever saw," he said sometimes, wiping the top of the can with the tail of his polo shirt, which was always crisply ironed.

"President Reagan is a good president!" he declared other times.

Still other times, he said nothing.

"He's an autistic," Myrtis told me. "That's different from AR-tistic."

Whatever it was, it was very mysterious. Donald was in his mid-thirties. He lived with his mother a few blocks away from our house. He spent his days riding his bike around town, occasionally mowing a lawn. He cut our grass on an irregular basis, showing up only on the President Reagan days. But he always came in to Pic-a-Sac for his Coke. Until one day, he didn't.

"Quick, climb up on my handlebars," Mom said, surprising me that night with a ride home on her bicycle.

I always walked to and from work at Pic-a-Sac, no matter the hour. It was only two blocks from our house.

"Donald's mom got icepicked," Mom said. "I hear he walked in and found her. Lying there. We've got to get home before anything happens to your sisters and brother."

It was 10 PM and the dark seemed darker than usual. We pushed the bike up onto the porch and ran into the house. Once inside, Mom locked all the doors and jammed the windows with wooden spoons. Hannah and Debbie were hiding in a closet, holding sleeping Michael.

I called Tommy.

"What do you want?" he said meanly.

Technically, we were broken up because Tommy thought I was in love with George, the black stock boy at Pic-a-Sac. Once when Tommy came by to visit me at work, George and I were back in the walk-in freezer, zapping each other with stickers from the price guns. One night after work, George and I went bowling. George paid for everything and gave me a ride home—well, not home, but to a big mansion I was pretending my family lived in.

"You sure this is your house?" George had said, clearly amused.

"Yes," I answered, jumping out of his car and running up the steps. I waved George away. "No need to wait 'til I get in—it's late, you just go on home now!"

Tommy had been so mad about the price-gun incident, I never told him about the bowling. I couldn't call George to come over and protect us because he thought I lived in that mansion.

"I'm scared," I said to Tommy. "There's an ice-pick maniac on the loose."

"Fuck it," he said. "I'm coming over."

Tommy got there twenty minutes later, swinging a long-barreled, Confederate-looking gun.

"I borrowed this from Vernon," he said, referring to his sister Noelle's husband. "It's a 420. It was Vernon's first rifle. He said y'all can keep it as long as you need to."

Tommy handed the rifle to my mother and dug some bullets out of his pocket. He showed us all how to load and aim it.

"Thanks, Tommy. Bye now," Mom said, dismissing him lest the neighbors see his car parked out front and think she'd allowed him to spend the night.

Mom moved the rocking chair over to the living room window. She sat down with the phone on one knee and the 420 on the other. Around three in the morning, both clattered to the floor. I found her shaking in front of the fireplace.

"I thought I saw something, and I pressed my nose up real close to the window, and there was a face looking *right back at me*," she huffed. "Dammit, I should've let Tommy spend the night on the couch."

We sat up together the rest of the night. When the sun came up, I ate a scrambled egg and went back to bed.

"Karen said we can sleep in her den tonight," Mom said, waking me up around noon.

"Who's Karen?" I asked.

"She's the one who lives two doors down, the one with three little kids?" Mom said brightly. "Her husband's out of town and she'd feel much better if we stayed over."

That evening, Mom, Hannah, Debbie, Michael, and I stood on the front steps of a stranger's house with our nightgowns and toothbrushes in a brown paper bag.

"Oh, hi!" Karen said when she opened the door. "Uhhhh, come on in—so sorry. I forgot your names."

"Thanks for having us all, Karen," Mom said. "I mean, it's just so scary, isn't it? And boy, did I get spooked last night!"

We sat on the sofa in the den while Karen fed her kids dinner in the kitchen.

"Are you sure you're not hungry?" she said. "There's plenty of spaghetti."

"No, we're fine!" Mom sang, shooting us a delirious look.

That night we lay on the sofas and floors, our stomachs grumbling.

"We'll just leave our toothbrushes here, okay?" Mom said the next morning.

"Okay," Karen said, her smile less forceful than yesterday.

We spent the next two nights in Karen's den. When we showed up the following evening, her husband answered the door.

"Oh, I see you're back," Mom said.

"Yes," he said.

"Are we going to sleep here again tonight?" Debbie asked.

"I want to watch cartoons on their big TV!" Michael chimed in.

"No," Mom said. "We're just here to pick up our toothbrushes."

✲ ✲ ✲

I requested a week off from Pic-a-Sac at the end of June. Dad was taking me, Hannah, Debbie, and Michael to Florida, the usual Orlando circuit.

"This'll be your last family vacation!" Dad said, as if he were praising me for some accomplishment. "You're getting in just under the wire."

In a couple of weeks, I'd be turning eighteen. I was often reminded that after this birthday, Dad would no longer pay monthly support for me, nor would I be entitled to any of the other fringe benefits of being a dependent child.

We did the Disney Universe, taking a turn around the new Epcot Center. I spent most of my time trying to keep Michael, now five, from stomping on the miniature frogs that had blanketed central Florida in plague-like numbers that summer. He did it with such gusto and no remorse. I wondered if he was a future serial killer.

On the fourth day in Florida, at Sea World, the veneer began to crack. Dad bought a frozen banana, thinking it was a banana-flavored popsicle. He bit through the chocolate shell, gagged and spat it out on the ground. He loved banana-*flavored* things, but he hated real bananas.

"Dammit," he said, wiping his tongue with a napkin. "Who wants the rest of this?"

After waiting in line for almost an hour, we entered Sea World's newest exhibit, the Shark Encounter. A dim maze of tanks holding dogfish and other small species led to the grand finale: a moving walkway that transported you through a giant shark tank via an eight-inch-thick Lucite tunnel. Ominous music played throughout the exhibit.

We stood on the walkway as ferocious-looking snaggletoothed sharks passed above us. Dappled, yellowish light filtered down through the tank.

"Sharks are not our enemy, though Hollywood movies want to make us think they are," a deep voice-over intimated. "Sharks are a natural, and necessary, part of the world's ecosystem, and do not attack man out of spite or hatred."

"Then why are you playing that spooky music?" Dad said, loudly. "Hypocrites!"

In the car on the way back to the hotel, Debbie rode shotgun. Hannah, Michael, and I sat in back. Under our breath, Hannah and I shared the secret shame of our mutual hunger and bargained over who would broach the topic with Dad.

"I asked him about lunch," I whispered.

"Okay, okay," Hannah mumbled. Tightening herself into a ball, with her chin tucked for safety, in a voice a little too loud and overly cheery, she said, "Sayyyyyy . . . what's for dinner?"

"WOULD YOU QUIT WITH YOUR PESTERING," Dad snarled. "All you kids want me for is my money. That's all you're after."

A remark in this vein usually shut everyone up. This time, I for one was tired of being a silent conspirator. Of all the kids in the car, I had the least to lose. I was getting cut off in two weeks. I felt brave enough to commit the gravest of sins: giving Dad some of my lip.

"Fathers," I said, slowly and deliberately, "are supposed to *want* to take care of their children."

Dad wrenched the car onto the shoulder of the highway and whirled around to face me. His eyes and neck veins burbled and bulged at intervals, like a nuclear reactor about to blow.

"WHY SHOULD I TAKE CARE OF YOU? YOU'LL JUST PUT ME IN A NURSING HOME WHEN I GET OLD."

Perhaps this would have been the ideal moment to point out that he'd put his own mother in a nursing home—yes—when she got old. I didn't have the nerve. The lifelong, almost inborn fear for my hide interceded, crushing my impulse to nurture Dad's sense of irony.

Irony was never his strong suit. One night, back when he still lived with us, he'd told Hannah: "Sure, honey, you can stay up late with me and your big sisters to watch that Holocaust movie."

Together we sat in the family room, watching *The Wall*, the story of a Jewish family's escape from the Warsaw Ghetto. Toward the end of the movie, the family is crawling through an underground sewer tunnel. The baby won't stop crying, no matter how much the mother tries to comfort it. Without a word, the father takes the baby from the mother. Looking into her eyes, he holds the baby tight against his chest until it quits crying. He suffocates it, so they won't be caught.

When the movie was over, Dad jumped out of his plaid recliner.

"What are you doing out of bed?" he demanded of Hannah.

Instinctively, Hannah started running down the hall toward her room. Dad ran after her. From the end of the hall, Becky and I watched him push her down onto the scratchy oatmeal carpet again and again, until he shoved her through her bedroom door.

Back on the side of the Florida highway, Dad and I glared into each other's eyes for I'm not sure how long. Long enough for me to imagine him lying on his deathbed, reaching for me with withered arms and rheumy eyes, begging for my forgiveness.

He was right.

I was an ingrate, sharper than a serpent's tooth. When the time came, I would drag him by a chain behind the car, dump him on the steps of the nursing home, peel out, and never look back.

Something inside me turned hard and cold.

"I'm hungry," Debbie said in the smallest of voices.

Dad turned around and looked at the road. He put the car into drive and pulled back out onto the road.

✲ ✲ ✲

"Big changes over at Pic-a-Sac while you were on vacation," Mom said. "Myrtis and L.W. are gone. Heard they got caught skimming off check-cashing fees. And you won't be seeing Red Man anymore. He got road-pizza'd out on Military Highway."

The whiff of death lingered over Pic-a-Sac, like Miss Havisham's parlor. The new assistant manager was a guy around forty-five, with reddish grey hair, named Brother. A long moustache hung like a polite curtain over his few remaining teeth.

Brother and I got along. He never enforced Mr. Ted's "Look Busy" policy, as he was a big ol' chatterbox himself. During lulls, he'd come over to my register, put his foot up on the bagging bench, his elbow on his knee, and his chin on his fist.

"I tell you what," he'd start, warming to his subject and ending with "ay-uh." Brother also did a pretty impressive hambone routine, slapping out catchy rhythms on his cheeks, hips, buttocks, and thighs.

One July afternoon when it was really dead, Brother stepped down from the manager's booth, locking the half-door. His keys retracted noisily back to his belt loop.

"Smoke break?" he said, tapping his pack and offering me one.

Everyone smoked at Pic-a-Sac, and we all smoked *in* Pic-a-Sac. Part of my job was changing the kitty litter in the ashtrays scattered throughout the store.

"Why you smoke menthols, Brother?" I asked as we lit up.

"I tell you what, Kool's got good promos," he said. "Got me these sunglasses, this lighter, free with every two packs you buy. Ay-uh."

He tipped his Kool visor at me.

"Understand you got a birthday comin' up next week," Brother said. "Wade got you little something."

Brother's son Wade worked back in the meat department. I'd only seen him once, when he came up front to get a shopping cart with a dead deer in it. One of our customers wanted him to butcher it. Wade stood there negotiating, covered with blood. All I really knew about him was that his father was named Brother.

"Oh, he shouldn't've gone to any trouble over me," I said, exhaling a minty cloud.

Brother made a motion toward the meat department with his hand, like a kitten batting a ball of yarn. Wade must have been watching through the tiny service window. He came bounding out of the swinging doors, cradling a bottle of Jean Naté perfume.

"Wade, you shouldn't have!" I said, meaning it.

I looked down at the bottle. Eau de cologne.

"Whatcha waiting for?" Brother said. "Put some on, girl."

Brother took the yellow bottle out of my hand and unscrewed the black ball top. He lifted it up to his nose.

"Mmmmm," he said, inhaling deeply.

I dotted the tiniest bit onto my index finger and touched the air behind my ears.

"Can't waste it!" I said, gagging at my new odor: Renuzit Air Freshener, Oil of Old Lady scent. "I want to save this up for special occasions."

"Like going to the movies with me Friday night?" Wade said.

I froze.

Pic-a-Sac whorled into a vortex of color. Wade, Brother, the deli ladies, and the rest of the gang pressed in close, chanting *One of us, one of us*. The curtain of Brother's moustache parted, and I shot jerkily into his gaping maw as if I were in a car entering the tunnel of a carnival spookhouse. The tracks ended and down I fell, down a rabbit hole of floating, chipped teeth in various stages of decay.

Bump! I landed on a spongy green tongue, face-to-face with my nemesis, Uncle Wiggly.

"Git ready for a lifetime o' menial labor!" he screeched. "Yeee-haw!"

Uncle Wiggly was an old baby tooth I still had up in the front of my mouth, the canine tooth to the left. He'd been loose for years. I'd tried everything to get him out: tongue, fingers, pliers, even some fishing line tied around him with the other end tied to a doorknob and a little sister slamming the door. Uncle Wiggly clung to my gum like a lamprey. I gave up; he won. I developed a crooked grin, sealing off one side of my lips to conceal him. I'd forgotten all about him until this moment.

"Thank you, Wade," I said. "But I'm already seeing someone."

"That's awright," said Wade. "I didn't want you anyway, bitch."

Soon I'd be up at college, LSU. The glittering metropolis of Baton Rouge would lie at my feet. To open doors, one needs teeth. Forget college—what about job interviews? What if Uncle Wiggly showed up at one of those?

"So tell me," the hiring person would say, "what are your qualifications for this job?"

"Well, ma'am, I—" I'd begin, only to be interrupted by a hillbilly voice coming from inside my mouth.

"Hey there, darlin'!" Uncle Wiggly would say. "Don't pay no mind to that girl! Look at ME. Get a load o' mah tricks! Watch me dangle, wrangle, and shingle-shangle, wiggle round, and almost fall down!"

Uncle Wiggly would slide out of my gum like a tusk and wink at the job interviewer. Uncle Wiggly would have eyes, a nose, and mouth of his own: a tooth with its own mouth.

"Heh heh. Nope, not comin out jest yet—gon' stay on rightchere! Hee hee! What more d'ya need to see? Hire me, fool! Hire me!"

"Er," I'd say, struggling to shove Uncle Wiggly back up into my gum, "you *do* provide dental coverage, right?"

Next thing you know, I'd be back at Pic-a-Sac, sawing antlers off a deer and splashing Jean Naté on my razor-burned neck.

One of us, one of us.

I made an appointment with Dr. Kennedy. He charged me only $20 for the consultation because I used to help his wife with her balloon bouquet business.

"First off, you'll need braces," Dr. Kennedy said, looking at my x-rays. "You'll have to see an orthodontist for that. There's a tooth up in your gum above the baby tooth, but it's sideways. Six months after you get the braces, the orthodontist will probably pull the baby tooth, then use a wire to guide the permanent tooth down into place."

The clock was ticking: I had one week before I turned eighteen.

"Dad, I need some dental work."

"Do what now?" he said, sitting across from me at Shoney's.

Dad had taken me out for lunch, sort of a Bon Voyage to Child Support Sendoff. He was already mad because I ordered the $4.99 baked fish off the menu, rather than getting the all-you-can-eat soup and salad bar for $3.99. Uncle Wiggly was tenacious, but I was afraid to eat crunchy foods. A rotten tooth was better than no tooth at all.

"See this?" I said, pulling up my lip with my pinky. "It's a baby tooth. It's been loose for ten years but hasn't come out 'cause the permanent tooth is stuck up in my gum somewhere, impacted-they-call-it, ay-uh."

Dad turned his head to the side and looked at me from the corner of one eye.

"Well," he said, "what do you want me to do about it?"

"The dentist says I need braces," I said, feeling like I was slogging up a staircase of wet cement. "I need you to pay for the braces. Please."

Dad turned his face back to look at me with both eyes.

"I done paid for all those years of allergy medicine, remember that?"

"Yes," I said.

"Inhalers, pills, scratch tests, antabottics," Dad said, tallying them up on his fingers.

"Yes," I said.

"I don't know, I invite you out to lunch, everything," he said, "just seems like all you do is ask for more, you're never satisfied, it's never enough."

"I'm sorry, Dad," I said, looking at the paprika sprinkled on top of my fish.

Dad squinted at me and slightly nodded, sizing me up.

"All right, then," he said.

Dad picked up his spoon and resumed eating his second bowl of tomato Florentine soup.

All right then?

Okay. Without dithering, for the first time in my life, I decided to take his words at face value.

✲ ✲ ✲

The very next day, I found a handsome orthodontist.

"In two years' time, you'll be as pretty as a pony," Dr. Boudreaux said, wheeling back in his chair after cementing the braces to my teeth. "First couple of weeks, you're going to feel some discomfort in that sweet little mouth of yours. Use the wax if you get sore spots. We'll see you next month to tighten the wires."

Exactly four weeks later, on a Friday in mid-August, I was back at Dr. Boudreaux's. I signed in at the front desk and took a seat in the waiting area. I opened an issue of *Family Circle*. Slowly, I became aware that the receptionists were whispering and pointing at me, nudging each other. Probably admiring my J. Crew t-shirt dress.

"Dr. Boudreaux will see you now," a nurse said, not making eye contact.

Instead of one of the plush examination rooms, she led me back to a small supply room, really just a large closet.

"Wait here," she said.

Dr. Boudreaux came in a few minutes later.

"Yes, well, Sandra, is it?" he said, suddenly walleyed. "Listen up. Your daddy never paid your bills, see, like you said he would, so we'll be needing a check today."

"There must be some mistake," I said, trying to convince myself there had been some mistake.

"Well, I hope so, Shawna," Dr. Boudreaux said. "We'll still need some payment today or we'll have to remove those braces, see, today."

I wanted to die of everything.

Dr. Boudreaux pulled out a pair of giant rusty wire cutters.

"All right, Sally," he said, snapping the wire cutters. "Let's go out back behind the Dumpsters in the parking lot and cut you out of those."

Well, not exactly. But that's what Uncle Wiggly wanted him to do:

"Bring it ON, fucker! Don't fence me in."

A desperate, long-slumbering part of me awoke.

What's the harm in letting us keep the braces? said my vestigial mooch. *I mean, what's he gonna do, put them in someone else's mouth? Like, gross.*

I shoved Mooch and Uncle Wiggly into an ornately carved armoire in the back of my brain.

"Dr. Boudreaux, sir," I said aloud, sweetly, hoping to catch more flies with honey, "my daddy hasn't been well since . . . the motorcycle accident. Please let me talk to him before we do something so rash."

His lips retracted from his perfect teeth like a convertible top.

"All right," he said. "Monday morning. *Cash down.*"

✲ ✲ ✲

"What should I say to get him to pay the two thousand dollars?" I asked Mom, hoping to ignite her past flair for prying money out of Dad.

It was the next morning, Saturday. Mom and I were in the kitchen. I was lying on the floor at her feet while she ironed some rich lady's sheets.

"I don't know," Mom sighed. "I can't get at his checkbook anymore, and even if I could, all the banks ask for photo identification now. If I were you, I'd let that orthodontist cut those braces off, or they'll sic a collection agency on you."

She dug her iron into the embroidered flange of a pillowcase and pushed the steam button.

Shahhhhhhhh.

The sound was like air being let out of a balloon. I stood up, then lost my balance and fell over, crashing into the kitchen sink.

"Sarah?" Mom said, setting down the iron.

"I'm okay," I said. "I just lost my equilibrium."

"You always had low blood pressure," Mom said, picking up the iron.

I held onto the back of a chair. I felt like Herman, my old friend Bitsy Marshall's dog. Years ago, bored one day, Bitsy and I got some scissors and snipped all the whiskers off Herman's right jowl. He took off running and slammed into a wall, then loped out the door and crashed into an azalea bush. Bitsy and I chased after him, all the way down to the pond on the golf course. He jumped in and swam out to the middle, thrashing around in an uneven circle, yalping nonstop. "Herman, Herman!" we called desperately from shore. Our cries made him flail and yalp even more. Finally, he made it back to shore. From then on, he was like that.

"I feel like Herman," I said to Mom.

"I'm not surprised," she said. "You've always been a little crazy."

"What?" I asked, semi-rejuvenated by any mention of me.

"When you were little, you'd run up with a book and grab my chin and say 'Read me read me read me!'" Mom said. "So I'd sit you down on my lap, and the second I started reading you'd be off and running, biting the nipple off your baby bottle. Getting a hold of you was like trying to get a hold of a bead of mercury from a broken thermometer. I know 'cause you broke two thermometers. Bit 'em clean in half."

"Great," I said, feeling weak again. "I probably have mercury poisoning on top of everything else."

"Then there was that time you stuck a key into the electrical socket," Mom said. "You were crying, saying 'Mommy, I got bizzed!' You learned your lesson, though. Never did it again."

"I need those braces," I said. "It's not fair. No one else has these problems."

"I never go to the dentist and I don't have your problems," Mom said, snapping a cotton sheet into a small, smooth rectangle.

I glanced at her teeth. The ones beyond the front four looked dubious.

"Look," she said, clamping down the corners of her mouth. "I'm supposed to babysit for Cookie Callahan tonight, but I have to do some research for a little project Michael and I are doing. Can you take the Cookie job?"

Cookie's kids were relatively well-behaved and went to bed early. The Callahan pantry was ample, bursting with Tostitos and Frito-Lay Bean Dip.

"All right, I'll do it," I said. "If I'm still alive by nightfall."

Mom unplugged the iron and opened a large reference book entitled *Artists of the Civil War.*

✵ ✵ ✵

Cookie Callahan was a trim, sporty woman. She ran the town's fancy boutique. Cookie enjoyed the finer things in life, like stirrup pants and Bud Light in a can. She lived with her well-dressed husband and children in a renovated old mansion. The ceiling in their foyer was emblazoned with a hot pink neon lightning bolt.

When Cookie and her husband got home later that night, instead of paying me and sending me on my way, she dragged a suitcase of Bud tallboys in from the garage fridge.

"Have a beer, Sarah," she said. "I know what hell my children can be."

Cookie really knew how to party. By the time we were on our fourth can of beer, I felt like we were old friends. I broke down and told her about Dad and the whole braces ordeal, and how now that I was eighteen I'm sure he wouldn't pay for them.

"Let it out," she said, scooting across the plush sofa and drawing me into her arms.

"Oh Cookie," I sobbed. "What am I gonna do?"

Cookie got very somber.

"Sarah, there's only one thing to do. You must sit your father down, look him the eye, and rustle up some tears—think of something sad, like the time you tried out for cheerleader and didn't make the squad—and scream, *Daddy, if you don't pay for these braces all my teeth are gonna fall out!!!*"

Cookie was watching herself in the panoramic mirror over the fireplace.

"Yeah, just like that, atta girl! That's how you wanna do it," she said, more to her reflection than to me. "That'll work real good!"

I closed my eyes and pictured myself saying that to my father. First, I superimposed Michael Landon's kindly face over Dad's.

"Pa," I said, "I reckon all my teeth are gonna fall out."

Michael Landon's face stayed frozen while his lips morphed into my father's.

"Do what now?" the lips said. "All your teeth're gonna fall out? Huh. Can't say that's a bad thing, if it keeps you out of my sunflower seeds."

Thanks, Cookie! Now I not only knew how abnormal it was to have a father who didn't care if all my teeth fell out, but I also felt sad that I hadn't made the cheerleading squad, when I'd practiced so hard! I brushed aside the niggling fact that I had once been on the kind of cheerleading squad you didn't need to practice and try out for, because you paid to be on it.

I mean, God, couldn't Cookie see it would be much easier if she would just spot me the cash?

"I just wish I could get the money from somewhere," I hinted, sniffling.

"Trust me," Cookie said, polishing off her beer. "No father can say no to a crying daughter."

"I just wish I had some sort of Fairy Godmother," I persisted, subtly. "Someone who looks young for her age."

Cookie Callahan didn't catch my drift. I stayed and drank a few more beers with her anyway. Couldn't hurt. She had a swimming pool in her backyard, and summer lasted until November.

"Well, I am ready to turn in," Cookie yawned. "Let me write you a check, honey. Is a dollar an hour all right?"

"Sure," I said, numb.

Cookie went out to the kitchen.

"Sorry, babe," she said, stumbling back into the living room several minutes later. "I can't seem to find my checkbook." She fished her wallet out of her purse and pulled it wide open. "Dang, I'm all out of cash. I'll have to owe you one."

I extricated myself from the deep, cushiony sofa.

"Now don't you worry, Sarah," Cookie said, leading me to the front door. "Everything's gonna be just fine, y'hear? Long as you make those tears look *real*."

✵ ✵ ✵

Sunday morning I got up and went to church with Mom.

"Forgive and forget," Father Mark said, launching into his sermon. "Sometimes we forget to forgive, don't we?"

Mom listened with her eyes closed, rapt. I slid a book from my purse into the church missal, quietly spreading its pages. Becky had loaned it to me: *The Unbearable Lightness of Being*.

The idea of eternal return is a mysterious one, I read, *and Nietzsche has often perplexed other philosophers with it: to think that everything recurs as we once experienced it, and that the recurrence itself recurs ad infinitum!*

Nietzsche? I think he was the guy Mom said was a Satanist. Reflexively, I looked skyward for a lightning bolt. Nothing. I read on.

Later that afternoon, I sat on the porch swing, waiting for Dad to bring Hannah, Debbie, and Michael back from their weekend at his place in Baton Rouge. Mentally rehearsing my last-ditch plea for dental asylum, I stared out at the broken grey asphalt of the road in front of our house. Lake Pontchartrain was that same grey, last time I went fishing with Dad.

It was right before my parents separated. Dad and I met up with Herbert Hebert before dawn at the boat launch in Slidell. Dad and Herbert put the boat in, and the three of us headed toward the twin span of bridges that crossed the lake to New Orleans.

I was thirteen. The only reason I still went fishing was to get a tan, and I was pissed because it was overcast that day. We fished for a couple of hours, catching nothing but black eels that looked like strips of rubber tire. When we threw them back, seemed like they got right back on our hooks.

A thunderstorm blew in. Our boat, a blue Lafitte skiff handmade by a fellow over in Manchac, tipped nearly perpendicular to each giant wave. I knew it was serious when I saw Dad do something I'd never seen him do before: put on a life jacket. He tried to get Herbert Hebert to put one on, too.

"Come on," Dad yelled. "Don't be such a damn fool!"

"No thanks," Herb smiled, shaking his head and turning the wheel of the boat to slice through a wall of water.

I crouched on the deck next to a big white Igloo cooler, fighting tears.

"Why won't he wear a life jacket?" I moaned.

"Guess he figures if his time's up, it's up," Dad said, watching Herb.

Well, what about *my* time? Not only were we all going to die, I was going to look pale and pasty in my casket.

Put on your fucking life jacket! I wanted to scream at Herb.

I wanted to scream. But I couldn't. All I could do was cower.

Mute.

Chicken.

The crackling sound of Dad's truck pulling up in front of the house made me jump. Hannah and Debbie and Michael climbed out, carrying the brown paper grocery bags containing their dirty clothes. I came down the porch steps, determined. They filed past me on their way in, radiating glumness.

"Where you been, stranger?" Dad asked me. "Still working over at the Piggly Wiggly?"

"Pic-a-Sac," I corrected him. Keeping my lip curled down over the braces made me sound like an old-time gangster.

"Getting ready for college, huh?" Dad said.

"Dad," I said.

He turned his head to the side in that familiar way: suspicious. Words tumbled out of my mouth.

"I got the braces like you said I could at Shoney's? But the orthodontist said he hadn't gotten any money from you? Maybe the bill got lost in the mail?" I said. Fucking weak. I switched gears. "Heh heh. They'll cut them off if you don't pay for them and I really need these braces, Dad." I swallowed, wondering if I should add, *Daddy, if you don't, all my teeth are gonna fall out!*

I couldn't say it. This time, I didn't want to lie.

"I got the bill," Dad said. "I never said I'd pay for braces."

"You did so say you would," I said. "At Shoney's, remember? You said 'all right.'"

"You went behind my back and signed up," Dad said. "You can pay for them yourself. Haven't you saved up enough from working at Winn-Dixie?"

A gloriously constructed diatribe against the travesty of minimum wage was forming in my brain, but I couldn't get the words lined up right.

"I'm not responsible for your medical bills anymore," Dad said. "You're eighteen."

He stalked off toward his truck.

"Dad, wait," I said, following him.

"You're a sneak," he said across the hood of the truck. "Just like your mother."

He got into the truck and slammed the door. He hitched up his hip, digging in his pocket for the keys. The side of his jaw pulsed.

I hate you! I wanted to scream.

Dad put his hands on the wheel and stared through the windshield. After a while, he stuck the keys in the ignition and put the truck into gear. It swerved off the shoulder back onto the grey road, spitting gravel.

Something snapped in my mouth. Without realizing it, I'd been pushing hard with my tongue on Uncle Wiggly. He was looser than ever, hanging by a string. I put my finger on him and pushed him back up into my gum, willing him to stay. A rotten tooth was better than no tooth at all.

I went out to the edge of the road and kicked at the gravel, unearthing a half-smoked cigarette butt. I bent and picked it up. More like two-thirds left than half, really. I brushed it off. Kool. I took a few steps toward Mom's porch, then stopped. In the living room window shone a lone candle-shaped bulb in a lamp with no shade.

I bowed my head and put the butt between my lips, then headed down the street. Somebody over at Pic-a-Sac was bound to have a light.

Acknowledgments

Thanks to my sibs, the Wolf Pack, for giving me this story to tell: to Becky—for reminding me of (i.e., correcting me on) the details; to R.; to Debbie, for being my eyes; and to Michael, for at least letting me *try* to turn him gay. This book wouldn't exist if the people in it hadn't, so thanks, Mom, for not believing in birth control. P.S. You are going to Heaven. Thanks to Dad, for understanding, and thanks to the state of Louisiana, the one place that nurtures eccentrics for the sheer joy of it.

A steamed little juicy bun of gratitude to David Rakoff, for his tirelessness and his appreciation of the finer things, like a walking beer in the Villa Borghese and a fold-out toilet. A large goblet of quality wine shall be raised to ye, O Cynthia Sweeney, because ye rule.

Thanks to my exceedingly rad agent, Erin Hosier, and to my editor, Amy Scheibe, and her unfailing instinct for how many pets should die in one book. Additional thanks to Jason Brantley, Nicole Caputo, Chris Greenberg, Beth Partin, and Kay Mariea.

Muchas gracias to Sonia Ramirez and Thelma Flores for taking such good care of my kids and our home while I wrote.

Big sloppy thanks to Paula Killen and Melanie Hutsell for the title and their unfailing support and humor. Arching ropes of thanks to Jill Soloway and Maggie Rowe for inviting me to Sit 'n' Spin, and to all the other Revolutionary Moths: Bernie Boscoe, Lisa Carver, Robin Goldwasser, Jessica

Kaminsky, Amanda Lasher, Jenny McPhee, Missie Noel, Jenifer Potts, Anne Preven, and Holiday Reinhorn.

Giant ham-handed clumps to Amy Sedaris and David Sedaris, for mistaking me for an actress, and a clump apiece to Jodi Lennon, Jackie Hoffman, Hugh Hamrick, Chuck Coggins, Paul Dinello, Stephen Colbert, and Greg Holliman. Seventy-nine bushels of awkward cuffs on the shoulder to the Upright Citizens Brigade—Matt Besser, Matt Walsh, Ian Roberts, and Amy Poehler. It was on their stage that the comedy summit of Pee Mountain was first reached.

Thanks to Dr. Steven Schwartz and his crack team of eyeball surgeons at UCLA, especially the one who laughed at my repeated, joking requests for medical marijuana. I owe my recovery to the Howard Stern radio show and the BBC World Service on KPCC.

Thanks also to Hilary Carlip and freshyarn.com, Emily Botein, Paget Brewster, Maria Curti, Rory Evans, L'Acajou, Jennifer Langkjaer, Chris Leavins, Barry Lieberman, Kathi Magnussen, Susan Sheu, Judy Stamper, Claudette Sutherland, Michael Scott, and Cynthia True.

Thankee kindly to all who contributed to or read *Thyrezine*, especially Laura Southwick, whom I think of and miss every day.

Thanks most of all, of all, to my husband, Andy Richter, for knowing when to protest the crazy and when to let it slide; and to W.O. and M.J., who make me follow through.